Introducing digital audio

Second edition

Ian R Sinclair

PC Publishing

PC Publishing
4 Brook Street
Tonbridge
Kent TN9 2PJ

This edition published 1992
First published 1988

ISBN 1 870775 22 8

British Library Cataloguing in Publication Data

Sinclair, Ian R. (Ian Robertson)
Introducing digital audio. – 2nd ed.
I. Title
621.3815

ISBN 1870775228

Phototypesetting by Scribe Design, Gillingham, Kent
Printed and bound in Great Britain by BPCC Wheatons Ltd, Exeter

Preface

The impact of digital methods on domestic sound reproduction has been as delayed in reaching us as it has been inevitable. Digital recording methods have existed for many years and have been familiar to the professional recording engineer, but the compact disc (CD) was the first device to succeed in bringing digital audio methods into the home, though Laservision (using analogue methods along with the optical disc recording system) had previously tried to break into the video market with limited success. Compact discs are now well established, to the extent that newspaper reviews of new issues concentrate more on CD releases than on releases on other media such as tape or LP (now referred to as 'black-disc'). In addition, several manufacturers are releasing new issues on CD only.

All this development has involved methods and circuits that are totally alien to the technician or keen amateur who has previously worked with audio circuits. The principles and practices of digital audio owe little or nothing to the traditional linear circuits of the past, and are much more comprehensible to the computer engineer than to the older generation of audio engineers. This situation has not been helped by the appearance of books that make the assumption that the reader has already mastered digital circuitry and also the mathematics of signal encoding. Since even the basics of digital circuitry are so totally unlike anything in linear circuits, the first two chapters of this book are devoted to digital principles and circuit devices rather than directly to digital audio.

This book is intended to bridge the gap of understanding for the technician and the keen amateur rather than for the professional audio engineer. In other words, the principles and methods will be explained, but the mathematical background and theory will be avoided other than to state the end-product. My aim is to show what is involved in the digital part of audio signals, particularly in the newer devices such as CD and DAT, rather than to go into details of sampling, error-correction and other esoteric points. It is important to note that digital audio methods at present account mainly for the CD player or the DAT player in the home, and at the time of writing the remainder of the audio chain of preamp and main amplifier remain firmly analogue with a few exceptions.

I am most grateful to all who have helped in this effort, particularly to Philips Ltd. and Sony (UK) Ltd. who have supplied vital information. I owe a special debt of gratitude to Phil Chapman who first suggested the book and has published it.

Ian Sinclair

Preface to the second edition

Since the first edition was published, many important developments have taken place in digital audio, and much more information is available. In particular, this book now deals with techniques such as oversampling and bitstream methods which were neglected earlier, and also looks at the Digital Compact Cassette, the Sony Mini Disc and the new domestic DAT players which are just appearing on the market.

Contents

1 One digit at a time

Digital electronics

The beginning of electronics was radio technology, and it is useful to remember that the first applications of radio to the transmission of messages used Morse code. A signal was (and still is) transmitted in Morse by switching the radio transmitter (in the early days, nothing much more than a spark-coil connected to an aerial and an earth) on and off, with the on time being either short (dot) or long (dash). This is a form of digital signal, because the essence of a digital signal is that it consists of two states only, on and off, with nothing in between having any significance. We would nowadays classify Morse as a digital pulse duration modulation, since the off time has no meaning, and the length of the on time determines either a dot or a dash being signalled, Figure 1.1. Morse code totally dominated radio until the development of transmitting valves in the 1920s, at which time it became possible to modulate radio frequencies with audio waveforms.

Even with the simplicity of Morse, problems can arise. Customarily, the dash has a duration of three times that of the dot, but

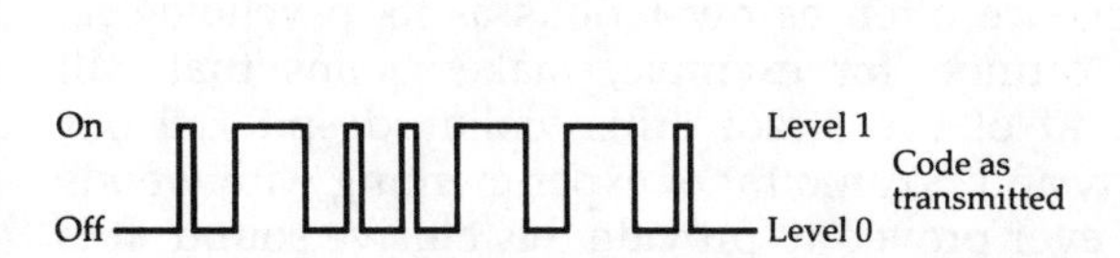

Figure 1.1 Morse was the first type of digital code. The significant part of the wave is the high (or 1) part, and the duration of this is used in a two-sense code.

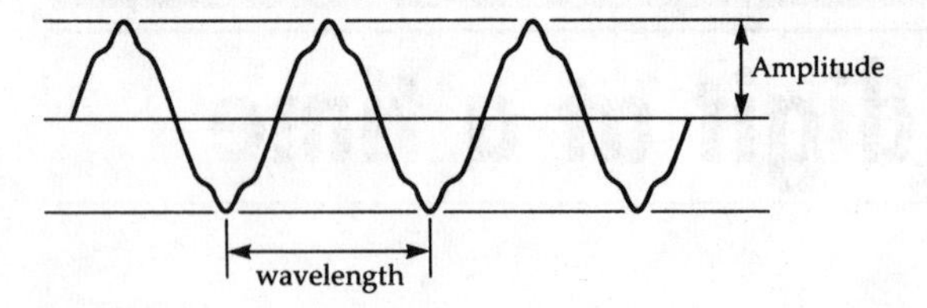

Figure 1.2 A typical audio waveform, simplified. A real audio waveform is always more complex, and never a sinewave.

when Morse is sent and received by human operators there can sometimes be confusion over long dots and short dashes. Even if the Morse is transmitted and read by machines, the effects of integration and other signal distortion can make it difficult at times to distinguish between dot and dash signals.

Recall at this point what we mean by an audio waveform, as might be represented by the graph of Figure 1.2. This shows a waveform, meaning that the pattern shape will repeat for several cycles, in which the voltage varies continually over the time of the cycle. A waveform like this will be, assuming that the system is a good one, a copy in electrical voltage terms of a sound-wave that was picked up by a microphone and which might be recorded or broadcast directly. The important point here is that the shape and frequency of the electrical wave is identical to the shape and frequency of the sound wave and the shape and frequency of the sound wave in turn is entirely responsible for the type of sound that we hear. The amplitude of the electrical wave, in volts, is proportional to the amplitude of the sound wave (in terms of air pressure), and is related to the loudness of the sound. The frequency of the electrical wave is the same as the frequency of the sound wave, and is responsible for our sensation of pitch. In short, the electrical wave is, or ought to be, a completely faithful copy of the sound wave and carries all the information of that wave.

This, in turn, is the start of all the problems that beset linear analogue recording systems. The sound wave is a complicated wave whose origins are often as mysterious as its psychological effect on us. We cannot, for example, make violins that will reproduce the sound of the Stradivarius violins despite all our modern technology, and no amount of experimenting with woods and varnishes has ever proved to provide this elusive sound. It is hardly surprising that the complexity of the sound from a full orchestra defies any straightforward analysis, and we have to add to all this complication the impact that the sound-waves make on our own ears. The ear itself is a most remarkable receiving instrument, and these that have escaped the attentions of the disco, the personal stereo and the ghetto-blaster are capable of detecting sound levels that are unmeasurably low and are yet capable of responding to subtle changes that theory tells us should be impossible to detect. The wonder is that any form of recording system could cope, and yet the history of recording is filled with witnesses saying that the sound is of wonderful quality (from the first horn gramophone) or of unparalleled fidelity (first electrical recording). In the early days, it was a wonder that it was done at all, and if the singing sounded anything like Galli-Curci in full sail, it seemed to be a miracle. A generation later, ears were sharper and senses more critical, and the noticeable improvements soon became the standards by which to judge further developments.

It has been so for each step in the improvement of recorded sound, from the flat wax disc to the LP record. Each time, we have been sure that perfection is within grasp, and within a few years we have become aware that it is not. Much of the problem lies in the recording medium itself. Consider what we are trying to achieve in cutting a master disc. A disc cutter is a mechanical cutting tool, a miniature chisel, that has to move in a pattern that exactly copies the change of voltage in the waveform so that it will cut a groove of that same shape in a disc. The sound-wave encounters no problems of this type – all it has to do is to move your ear-drum, which is a considerably lighter device, and with no requirement to cut grooves in your skull. No mechanical device can easily be vibrated in the pattern of a sound wave, and it is hardly surprising that the cutters of recording lathes require enormous amplifier power in order to achieve results – 500 to 1000 watts are the kind of figures we are talking about for each channel. The original discs have then to be copied by moulding methods, adding yet another source of distortion. Finally, the copied recording is played by lowering a stylus into the groove, spinning the disc and expecting the shape of the groove to vibrate the stylus in exactly the pattern of the groove shape. This would be difficult enough if only the lightweight stylus had to be moved, but the stylus has to be connected to some form of electrical transducer, the pickup, and this in turn means that movement is not so easy, and perfect reproduction of the waveform is that much more difficult.

You might imagine that magnetic tape recording offered a solution, and to some extent it can for studio work. The problem

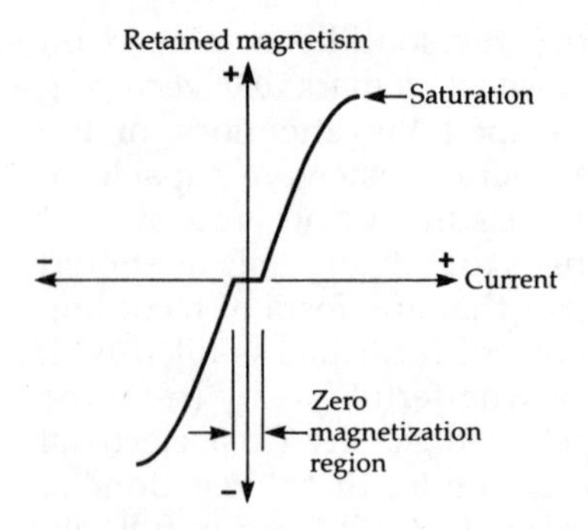

Figure 1.3 A typical tape magnetization characteristic. Most of the shape is curved, and there is a region in which small magnetizing signals leave no trace at all on the tape.

of all magnetic recording, however, is that the whole process is inherently non-linear. If you plot the intensity of retained magnetism for a magnetic material against the current through a coil that causes the magnetization, you arrive at a severely curved graph shape (Figure 1.3) which is not even continuous. For very small amounts of coil current, the tape is not magnetized at all, and for large amounts of current the tape saturates, meaning that it becomes magnetized so that the range of amplitudes that can be recorded is severely limited. The range of frequencies that can be recorded is limited by the combination of tape speed and recording head gap. In addition, the recorded tape has a noticeable noise level that can be heard clearly during passages of soft music and is very obtrusive when narrow strips of tape are used. All of these problems can, of course, be attacked by using ultrasonic bias, wide tapes, high tape speeds and volume compression, and tape mastering is used extensively, though only too often for the purpose of doctoring recordings rather than for reasons of fidelity.

Even the transistor, almost universally used for amplification, is by no means a linear device, and the history of hi-fi is virtually the history of circuit techniques that have been devised to get around the inherent non-linearity of, at first, valves and subsequently transistors and ICs. We appear to find a new cause of distortion about each ten years and take about ten years to find a palliative, so that the audio scene has never been static, and recording engineers have been known to wonder if some critics would be satisfied with the original sound, let alone a recording. In general, though, electrical distortions are easier to deal with than mechanical distortions, and the weakest links in any sound reproducing system have been recognised as the recording process itself and the loudspeaker. The use of digital methods is the first step in a fresh look at the whole system, a complete rethink of the recording process.

Digital signals

A digital signal is one in which a change of voltage, and the time at which it occurs, are of more importance than the precise size of the change. Nothing is conveyed by the shape of a digital waveform, nor even by its precise amplitude. All of the waveforms in digital circuits are steep-sided, pulses or square waves, and it is the change of voltage at either side of the pulse that is significant. For that reason, the voltages of digital signals are not referred to directly, only as 1 and 0 values. The important feature of a digital signal is that each change is between just two voltage levels, typically 0V and +5V, and that these levels need not be precise. In this example, the '1' level can be anything from 2.4V to 5.2V, and the '0' level anything from 0V to 0.8V. By using 0 and 1 in place of the actual voltages, we make it clear that digital electronics is about numbers, not waveforms or amplitudes. In digital circuits, the pulses will be used to make or break connections, using circuits called gates, or they will be counted. In neither case is the shape of the pulse important other than in the requirement that the change of voltage between the levels must be as fast as possible. In neither case is the amplitude of the pulses important provided that each pulse is between some voltage that is acceptable as a 0 level and another voltage that will be acceptable as a 1 level.

The importance of using just two digits, 0 and 1, is that this is ideally suited to electronic devices and also to tape recording. A transistor, whether bipolar or FET, can be switched fully on or fully off by a suitable voltage at its input (base or gate), and these two states can be ensured easily, much more easily than any other states because no form of bias is needed. By using just these two states, then, we can avoid the kind of errors that would arise if we tried to make a transistor operate with, say, ten levels of voltage between the two extreme voltage levels of cut-off and saturation. By using only two levels, the possibility of mistakes is made very much less. There are, for example, no problems about bias since the only requirement is that the device shall be biased either fully on or fully off. There are no problems about linearity, because nothing is being amplified, the output signal from a digital circuit will normally be of the same amplitude as the input signal. The only snag is that any counting that we do has to be in terms of

Place No.	0	1	2	3	4	5	6	7	8	9	10	11	12
Denary	1	2	4	8	16	32	64	128	256	512	1024	2048	4096

Place No.	13	14	15
Denary	8192	16384	32768

To convert a binary number into denary, find the denary figure for the place number of each 1, then add the denary numbers. Remember that the place numbers are counted from the right hand side, starting with zero.

For example, the number 10010110 contains a 1 in positions 1, 2, 4 and 7. These place numbers correspond to 2, 4, 16 and 128 respectively, so that the denary equivalent is 150.

Figure 1.4 A table of powers of two, and how the table can be used in converting from 8–4–2–1 binary code into denary numbers.

Denary to binary conversion is done on paper by dividing the denary number by 2, noting the remainder, and then repeating the process on the result of division until the final remainder (which is always 1) is obtained. The binary number consists of the remainder digits read from the bottom up.

Example: Convert the number 583 to binary.

583/2 = 291 remainder 1
291/2 = 145 remainder 1
145/2 = 72 remainder 1
72/2 = 36 remainder 0
36/2 = 18 remainder 0
18/2 = 9 remainder 0
9/2 = 4 remainder 1
4/2 = 2 remainder 0
2/2 = 1 remainder 0
1/2 = 0 remainder 1

Binary number is 1001000111

Figure 1.5 How denary numbers can be converted to binary on paper.

only two digits, 0 and 1. For some applications, this is of no importance because counting may not be involved. If you use a digital circuit, for example, to control two gas valves, then the outputs of the digital circuit will turn each valve either fully on or fully off. The action is one of control only, not of counting. On the other hand, we might want to turn one valve on after two pulses at an input, and the other valve on after four pulses at the same input, and this action is quite definitely one of counting.

Counting with only two digits means using a scale of two. There are many types of scales that can be used, as we shall see later, but the most important one is the 8–4–2–1 scale. There's nothing particularly difficult about this, because numbers in this scale, often called simply the binary scale, are written in the same way as ordinary numbers (denary numbers or scale-of-ten numbers). As with denary numbers, the position of a digit in a number is important. For example, the denary number 362 means three hundreds, six tens and two units. The positions represent powers of 10, with the right hand position (or least-significant position) for units, the next for tens, the next for hundreds (ten squared), then thousands (ten cubed) and so on. For a scale of two, the same scheme is followed. In this case, however, the positions are for units, twos, fours (two squared), eights (two cubed) and so on. The table Figure 1.4 shows powers of two and how a binary number can be converted into denary form. Figure 1.5 shows the conversion in the opposite direction. The maximum number that can be expressed in a scale of two depends, as it does in any numbering system, on how many digits are used. If we allowed only two binary digits (or bits) then we would be counting with a number range of 0 to 3 only. By using 8 digits (called one byte), we can cope with numbers from 0 to 255, and by using sixteen bits (called one word) we can make use of a number range from 0 to 65535. The number of bits that can be used is therefore an important feature of any digital system that involves counting, and this is something that we shall return to later.

Digital circuits are switching circuits, and the important feature is fast switching between the two possible voltage levels. Most digital circuits would require a huge number of transistors to construct in discrete form, so that digital circuits make use of ICs almost exclusively. These ICs can make use of either bipolar transistors in integrated form, or of MOSFETs, and both types are extensively used. MOSFET types are used extensively in computing as memory circuits and in the form of microprocessors. The bipolar types are used where higher operating speeds and larger currents are required, and these applications occur both in computer circuits and in industrial controllers. The digital ICs that are used in digital audio circuits are almost invariably purpose-designed ICs rather than the types of over-the-counter chips that would be used in other types of digital circuitry.

As it happens, the development of digital ICs has had a longer history than that for analogue devices. When ICs could first be produced, the manufacturing of analogue devices was extremely difficult because of the difficulty of ensuring correct bias and the problems of power dissipation. Digital IC circuits, using transistors that were either fully off or fully on, presented no bias problems and had much lower dissipations. In addition, circuits were soon developed that reduced dissipation still further by eliminating the need for resistors on the chip. Digital ICs therefore had a head start as far as design and production was concerned, and because they were immediately put into use, the development of new versions was proceeding before analogue ICs made any sort of impact on the market.

Given, then, the advantages of digital signals as far as the use of transistors is concerned, what are the wider advantages for the recording of signals? The most obvious advantage relates to tape or any other magnetic recording. Instead of expecting the magnetization of the tape to reproduce the voltage of a signal, the tape magnetization will be either maximum in one direction or maximum in the other. This is a technique called 'saturation recording' for which the characteristics of most magnetic recording materials are ideally suited. The precise amount of magnetization is no longer important, only its direction. This, incidentally, makes it possible to design recording and replay heads rather differently so that a greater number of signals can be packed into a given length of track on the tape. Since the precise amount of magnetization is not important, linearity problems disappear.

Tape noise consists of signals that are far too small to register, so that they have no effect at all on the digitally recorded signals. This also makes tapes easier to copy, because there is no degradation of the signals caused by copying noise, as there always is when conventional analogue recorded tapes are copied. Since linearity and noise are the two main problems of any tape (or other magnetic) recording system it is hardly surprising that major recording studios have rushed to change over to digital tape mastering. The surprising thing is that it has been so late in arriving on the domestic scene, because the technology has been around for long enough, as long as that of videotape recording.

The advantages that apply to digital recording with tape apply even more forcefully to discs. The accepted standard method of placing a digital signal on to a flat plastic disc is to record each digit bit as a tiny dimple on the otherwise flat surface of the disc, and interpret a digital 1 as the change of reflection of a laser beam. Once again, the exact size of the dimple is unimportant as long as it can be read by the beam, and only the number of dimples is used to carry signals. We shall see later that the process is by no means so simple as this would indicate, and the CD is a much more complicated and elaborate system than the tape system (DAT). The basic principles, however, are simple enough, and they make the system immune from the problems of the LP disc. There is no mechanical cutter, because the dimples have been produced by a laser beam which has no mass to shift and is simply switched on and off by the digital signals. At the replay end of the process, another (lower-power) laser beam will read the pattern of dimples and once again this is a process which does not require any mechanical movement of a stylus or any pickup mechanism, and no contact with the disc itself.

As with magnetic systems, there is no problem of linearity, because it is only the number of dimples rather than their shape and size that counts. Noise exists only in the form of a miscount of the dimples, and as we shall see there are methods that can reduce this to a negligible amount. Copying of a disc is not so easy as the moulding process for LPs, and mass production requires enormous capital investment and costly inspection processes, all of which are threatened by the advent of digital tape systems. A CD copy, however, is much less easily damaged than its LP counterpart, and even discs that look as if they had been used to patch the M1 will play with no noticeable effects on the quality of the sound — though such a disc will sometimes skip a track. The relative immunity to wear is a very strong point, because this allows CDs to recoup their high costs when they are used in jukebox applications or for the inescapable deluge of recorded sound in restaurants and other public places. This has also allowed CD libraries to flourish, because each disc can earn its keep in loan fees and can afterwards be sold at a good price because it can still be played with no noticeable deterioration in quality.

Paying the price

No advantages are ever obtained without paying some sort of price, and the price to be paid for the advantages of digital recording and reproduction consists of the problems of converting between analogue and digital signal systems, and the increased rate of processing of data. To start with, the sound wave is not a digital signal, so that its electrical counterpart must be converted into digital form. This must be done at some stage where the electrical signal is of reasonable amplitude, several volts, so that any noise that is caused will be negligible in comparison to the signal amplitude. That in itself is no great problem, but the nature of the conversion is. What we have to do is to represent each part of the wave by a number whose value is proportional to the voltage of the waveform at that point. This involves us right away into the two main problems of digital systems, resolution and response time.

To see just how much of a problem this is, imagine a system that used numbers −2 to +2 only, used on a signal of 4V total peak-to-peak amplitude. If this were used to code a sinewave as in Figure 1.6(a) then since no half-digits can exist, any level between −0.5V and +0.5V would be coded as 0, any signal between +0.5V and +1.5V as 1 and so on, using ordinary denary numbers rather than binary numbers to make the principle clearer. In other words, each part of the wave is represented by an integer (whole) number

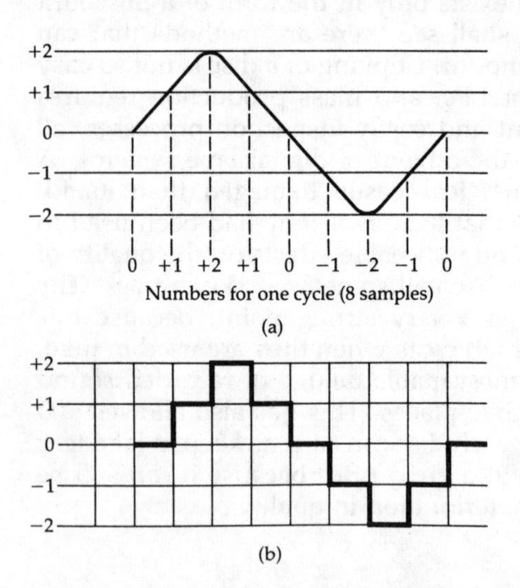

Figure 1.6 (a) Coding an analogue signal into five levels (zero is counted as one level), so that each part can be represented as an integer number (no fractions). The replay (b) of this wave shows the block shape due to the five level quantization.

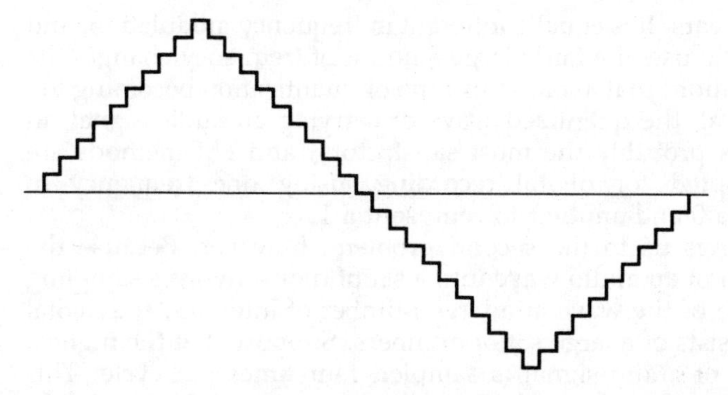

Figure 1.7 The result of a quantization with a much larger number of levels. The waveshape in this example is triangular because this makes it easier to show the effect of the quantization.

between −2 and +2, and if we plotted these numbers on the same graph scale then the result would look as in Figure 1.6(b). This is a 'block' shape of wave, but recognisably a wave which if heavily smoothed would be something like the original one. We could say that this is a five-level quantization of the wave, meaning that the infinite number of voltage levels of the original wave have been reduced to just five separate levels. This is a very crude quantization, and the shape of a wave that has been quantized to a larger number of levels is shown in Figure 1.7. The larger the number of levels, the closer the wave comes to its original pattern, though we are cheating in a sense by using a sinewave as an illustration, since this is the simplest type of wave to convert in each direction. Nevertheless, it is clear that the greater the number of levels that can be expressed as different numbers then the better is the fidelity of the sample.

In case you feel that all this is a gross distortion of a wave, consider what happens when an audio wave of 10kHz is transmitted by radio, using a carrier wave of 500kHz. One audio wave will occupy the time of 50 radio waves, which means in effect that the shape of the audio wave is represented by the amplitudes of the peaks of 50 radio waves, a 50 level quantization. You might also like to consider what sort of quantization is involved when an analogue tape system uses a bias frequency of only 110 kHz, as many do. The idea of carrying an audio wave by making use of samples is not in any way new, and is inherent in amplitude modulation radio systems which were considered reasonably good for many years. It is equally inherent in frequency modulation, and it is only the use of a fairly large amount of frequency change (the peak deviation) that avoids this type of quantization becoming too crude. Of all the quantized ways of carrying an audio signal, in fact, FM is probably the most satisfactory, and FM methods are often adopted for digital recording, using one frequency to represent a 0 and another to represent a 1.

This brings us to the second problem, however. Because the conversion of an audio wave into a set of digits involves sampling the voltage of the wave at a large number of intervals, the digital signal consists of a large set of numbers. Suppose that the highest frequency of audio signal is sampled four times per cycle. This would mean that the highest audio frequency of 20kHz would require a sampling rate of 80 kHz. This is not exactly an easy frequency to record even if it were in the form of a sinewave, and the whole point of digital waveforms is that they are not sinewaves but steep-sided pulses which are considerably more difficult to record. From this alone, it's not difficult to see that digital recording must involve rather more than analogue recording.

The next point is the form of the numbers. We have seen already that numbers are used in binary form in order to allow for the use of only the two values of 0 and 1. The binary code that has been illustrated in this chapter is called 8–4–2–1 binary, because the position of a digit represents the powers of two that follow this type of sequence. There are, however, other ways of representing numbers in terms of 0 and 1, and codes such as the Gray code (Figure 1.8) and the Excess-3 code are used for some commercial processes. The advantage of the 8–4–2–1 system is that both coding and decoding are relatively simple. Whatever method is used, however, we cannot get away from the size of a binary number. It is generally agreed that modern digital audio should use a sixteen-bit number to represent each wave amplitude, so that the wave amplitude can be any of up to 65536 values. For each sample that we take of a wave, then, we have to record 16 digital signals, 0 or 1, and all 16 will be needed in order to reconstitute the original wave.

This is the point on which so many attempts to achieve digital coding of audio have foundered in the past. As so often happens, the problems could be more easily solved using tape methods, because it would be quite feasible to make a sixteen-track tape recorder using wide tape and to use each channel for one particular bit in a number. This is, in fact, the method that can be used for digital mastering where tape size is not a problem, but the

Denary	8-4-2-1 Code	Gray Code
0	0000	0000
1	0001	0001
2	0010	0011
3	0011	0010
4	0100	0110
5	0101	0111
6	0110	0101
7	0111	0100
8	1000	1100
9	1001	1101
10	1010	1111
11	1011	1110
12	1100	1010
13	1101	1011
14	1110	1001
15	1111	1000

Gray scale converters use four bits only, because the conversion between four-bit binary allows binary-coded denary (BCD) systems to be used. If true 8-4-2-1 binary is needed, then converters between BCD and 8-4-2-1 can be used also.

Figure 1.8 The Gray code, which is used in preference to the 8-4-2-1 type for many industrial control applications. The main advantage is that only one digit changes at a time as the count progresses. IC converters can be obtained for four-bit Gray code to or from four-bit binary code.

disadvantage here is that for original recordings some 16 to 32 separate music tracks will be needed. If each of these were to consist of sixteen digital tracks the recorder would, to put it mildly, be rather overloaded. Since there is no possibility of creating or using a sixteen-track disc, the attractively simple idea of using one track per digital bit has to be put aside. The alternative is serial transmission and recording.

Serial means one after another. For 16 bits of a binary number, serial transmission means that the bits are transmitted in a stream of sixteen separate signals of 0 or 1, rather in the form of sixteen separate signals on sixteen channels at once. Now if the signals are samples taken at the rate of 60kHz, and each signal requires sixteen bits to be sent out, then the rate of sending digital signals is 16 × 60 kHz, which is 960 kHz, well beyond the rates with which ordinary tape or disc systems can cope. As it happens, we can get away with slower sampling rates, as we shall see, but this doesn't offer much relief because there are further problems. When a parallel system is used, with one channel for each bit, there is no problem of identifying a number, because the bits are present at the same time on the sixteen different channels. When bits are sent one at a time, though, how do you know which bits belong to which number? Can you be sure that a bit is the last bit of one number or is it the first bit of the next number? The point is very important because when the 8–4–2–1 system is used, a 1 as the most important bit means a value of 32768, but a 1 as the least important bit means just 1, Figure 1.9. The difference in terms of signal amplitudes is enormous, which is why codes other than the 8–4–2–1 type are used industrially. The 8–4–2–1 code is used mainly in computing because of the ease with which arithmetical operations can be carried out on numbers that use this code.

Even if we assume that the groups of sixteen bits can be counted out perfectly, what happens if one bit is missed or mistaken? At a frequency of 1 MHz or so it would be hopelessly optimistic to assume that a bit might not be lost or changed. There are tape dropouts and dropins to consider, and discs cannot have perfect surfaces. At such a density of data, faults are inevitable, and some methods must be used to ensure that the groups of sixteen bits, called 'words' remain correctly gathered together. Whatever method is used must not compromise the rate at which the numbers are transmitted, however, because this is the sampling rate and it must remain fixed. Fortunately, the problems are not new nor unique to audio; they have existed for a long time and been tackled by the designers of computer systems. A look at how

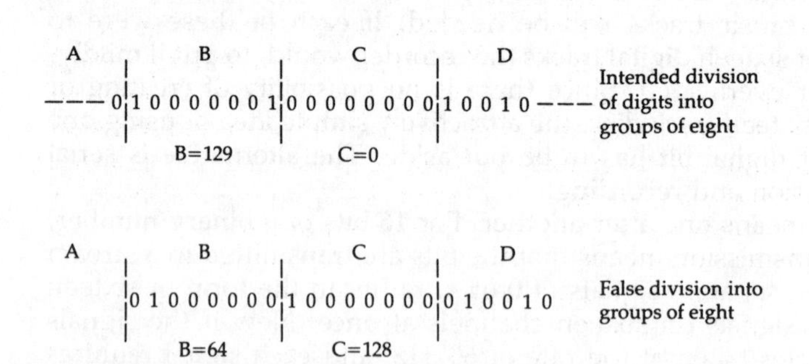

Figure 1.9 An error caused by a missing digit in a serial transmission could cause each following byte to be wrongly read.

these problems are tackled in simple computer systems gives a few clues as to how the designers of audio digital systems went about their task.

To start with, when computers transmit data serially, the word that is transmitted is not just the group of digits that is used for coding a number. For historic reasons, computers transmit in units of eight bits, called a byte, rather than in 16-bit words, but the principles are equally valid. When a byte is transmitted over a serial link, using what is called asynchronous methods, it is preceded by one 'start bit' and followed by one or two (according to the system that is used) 'stop bits'. Since the use of two stop bits is very common, we'll stick to the example of one start bit, eight number bits and two stop bits. The start bit is a 0 and the stop bits are 1's, so that each group of eleven bits that are sent will start with a 0 and end with two 1's. The receiving circuits will place each group of eleven bits into a temporary store and check for these start and stop bits being correct. If they are not, then the digits as they come in are shifted along until the pattern is now correct – the process is shown in outline in Figure 1.10. This means that an incorrect bit will cause loss of data, because it may need several attempts to find that the pattern fits again, but it will not result in every byte that follows being incorrect, as would happen if the start and stop bits were not used.

The use of start and stop bits is one method of checking the accuracy of digital transmissions, and it is remarkably successful, but it is just one of a number of methods. In conjunction with the use of start and stop bits, many computer systems also use what is known as parity, a method of detecting one-bit errors. In a group of eight bits, only seven are normally used to carry data and the eighth is spare. This redundant bit is made to carry a checking signal, which is of a very simple type. We'll illustrate how it works with an example of what is termed even parity. Even parity means that the number of 1's in a group of eight shall always be even. If the number is odd, then there has been an error in transmission and a computer system may be able to make the transmitting equipment try again. When each byte is sent the number of 1's is counted. If this number is even, then the redundant bit is left as a 0, but if the number is odd, then the redundant bit is made a 1, so that the group of eight now contains an even number of 1's. At the receiver, all that is normally done is to check for the number of 1's being even, and no attempt is made to find which bit is at fault if an error is detected. The redundant bit is not used for any purpose other than making the total number even.

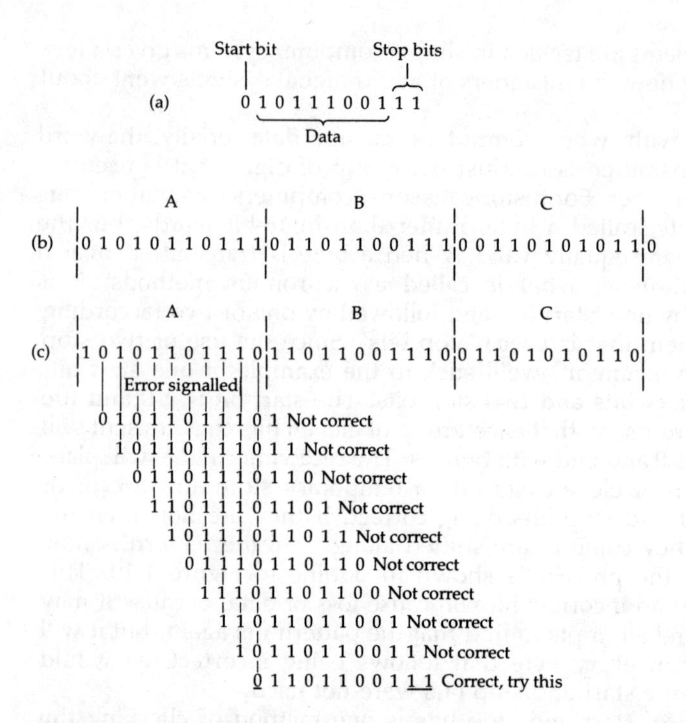

Figure 1.10 (a) Using one start and two stop bits for each group of eight data bits so as to make synchronization easier for serial systems. (b) The form of a three-byte signal with start and stop bits added. (c) How an error in the first bit will be discarded. The bits are read in groups of eleven, shifted by one bit each time. Data is not used until the pattern of a zero, eight data bits and two 1's can be found again.

Parity, used in this way, is a very simple system indeed, and if two bits in a byte are in error it is possible that the parity could be correct though the transmitted data was not. In addition, the parity bit itself might be the one that was affected by the error so that the data is signalled as being faulty even though it is perfect. Nevertheless parity, like start bits and stop bits, works remarkably well and allows large masses of computer data to be transmitted over serial lines at reasonably fast rates. What is a reasonably fast rate for a computer is not, however, very brilliant for audio, and even for the less-demanding types of computing purposes, the use of parity is not really good enough, and much better methods have been devised. The rates of sending bits serially for computing purposes range from the abysmally slow 110 bits per second to the rather unreliable 19,600 bits per second. Even this fast rate is very slow by the standards that we have been talking about, so it's obvious that something rather better is needed for audio information. This is something that we shall be looking further into in Chapter 3 and also in subsequent chapters.

All in all, then, you can see that the advantages that digital coding of audio signals can deliver is not obtained easily, whether we work with tape or with disc. The rate of transmission of data is enormous, as is the bandwidth required, and the error-detecting methods must be very much better and work very much faster than those needed for the familiar personal computers that are used to such a large extent today. That the whole business should have been solved so satisfactorily as to permit mass production is very satisfying, and even more satisfying is the point that there is just one world-wide CD standard, not the furiously competing systems that have made video recording such a problem for the consumer.

2 Digital devices

Gates

The gate, or logic gate to give it its full title, is the building block component of digital circuits, rather in the way that the single amplifier stage is the building block for conventional audio circuits. Gates for digital use come in two main types, called AND and OR, and the action is rather more strictly defined than that of a gate in a linear circuit. Our idea of a linear gate is a circuit that will pass or block a main signal in response to a gating signal on another input, as Figure 2.1 suggests. In this respect, a gate in a linear

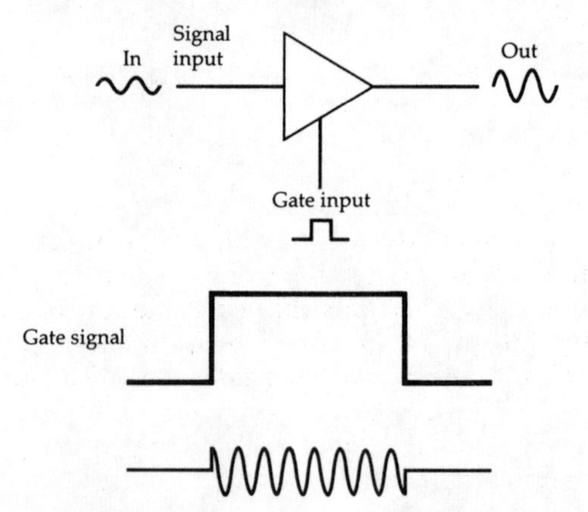

Figure 2.1 The action of a linear gating circuit. The main input and outputs will be analogue signals, with a square-sided signal used to switch the gate on and off. The waveshape of the analogue signal is preserved, and the only gate law is that the gate is open for a high voltage at the gating input and closed for a low gating input.

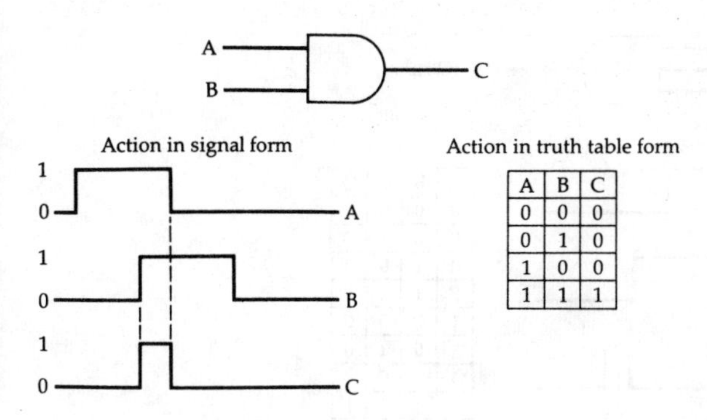

A	B	C
0	0	0
0	1	0
1	0	0
1	1	1

Figure 2.2 The action of the AND digital gate. Inputs and outputs are digital signals of about the same amplitude, and the output depends on the combination of inputs. In this case, the output is 1 only while both inputs are at 1. The action has been shown both in signal form and in the more common truth-table form.

circuit is no more than an on/off switch for signals. Though the digital logic gate can also be used in this way, its distinguishing feature is that the conditions that determine the output of the gate obey a set of rules that can be summarised in the form of a truth table, and which are affected by all the inputs to the gate, unlike the analogue gate in which one input is of a signal and the other is a gating waveform which determines whether the signal is passed or not.

Figure 2.2 shows a typical truth table along with the conventional symbol for one gate type, the AND gate. These symbols are taken from the international set that are used by manufacturers and users. The important point is that the gate illustrated here has two inputs, both of these inputs are digital signals, and the signal that you get at the output of the gate depends on both of the input signals. Since a digital signal can have two values only, 0 or 1, the table shows only these signals for each possible combination of inputs. As you can see, the output is 0 for three combinations of inputs, and only for the fourth does the output change to 1. This is when input A **and** input B are together at level 1, and it's for this reason that the gate is called an AND gate. If three inputs are to be used, the truth table and symbol are as shown in Figure 2.3. There is still only one condition that results in a 1 at the output — when all inputs are at 1. No matter how many inputs an AND gate is designed to use, then, there is only one combination of inputs that produces a 1 at the output, and that's when all of the inputs are at level 1.

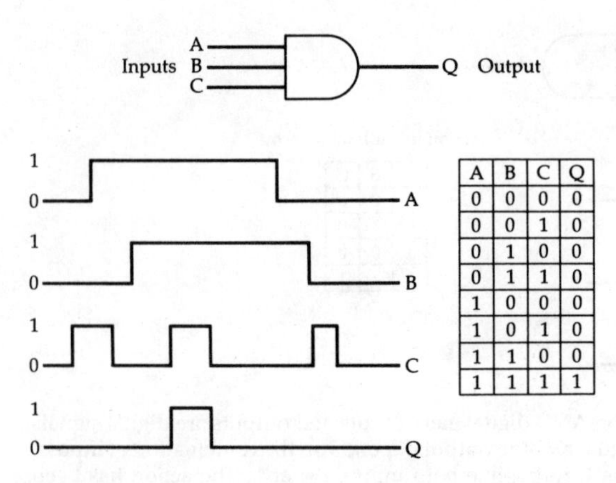

A	B	C	Q
0	0	0	0
0	0	1	0
0	1	0	0
0	1	1	0
1	0	0	0
1	0	1	0
1	1	0	0
1	1	1	1

Figure 2.3 The action of a three-input AND gate shown both in signal form and in truth table form.

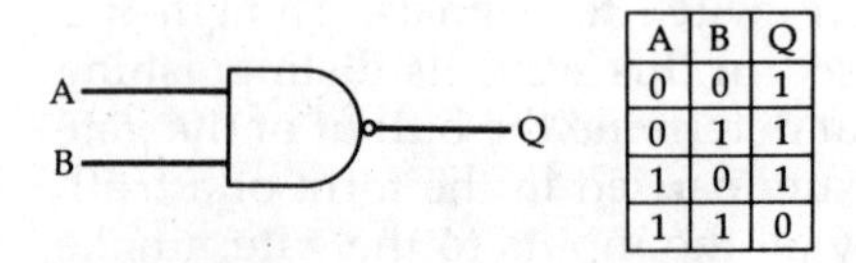

A	B	Q
0	0	1
0	1	1
1	0	1
1	1	0

Figure 2.4 The symbol and truth table for the NAND gate, which is the equivalent of an AND gate followed by an inverter.

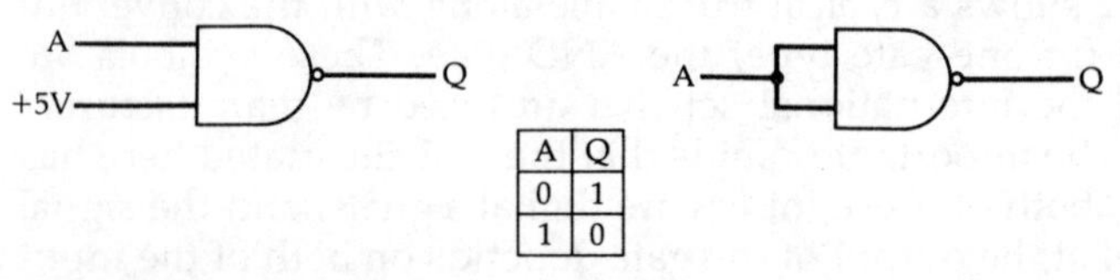

A	Q
0	1
1	0

Figure 2.5 How the NAND gate can be used as an inverter. Any logic system can be made up from NAND gates (or from NOR gates) alone, so that this gate is a fundamental building block of digital ICs.

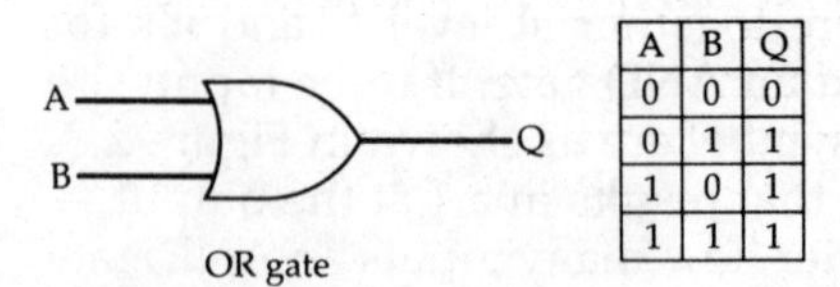

A	B	Q
0	0	0
0	1	1
1	0	1
1	1	1

Figure 2.6 The action of the OR gate, illustrated here for two inputs. Note that this is not our everyday idea of OR, because it includes the case where both inputs are at 1.

As it happens, it's very much easier to manufacture an inverted form of this gate, known as the NAND gate, and its truth table and symbol are shown in Figure 2.4. The small circle shown here at the output of the symbol means inversion, and the N in NAND means NOT. The output is 1 except when inputs A AND B are 1, when the output becomes 0. The value of the NAND gate is that it can be used for more than one purpose. Suppose, for example, that you connect the gate in either of the two methods shown in Figure 2.5, using earth to provide logic level 0 and the supply voltage to provide logic level 1. Either way round, the gate acts so as to invert the signal at its input, supplying a 0 if the input is 1 or a 1 if the input is 0. This is just one example of the versatility of this type of gate, and for a full explanation of how it can be used to provide any type of gate action that you might want you will have to consult a book such as my *Digital Logic Gates and Flip-Flops* (also from PC Publishing) because we haven't space to go into all the byways of gate circuits here.

The other main gate type is the OR-gate, and the symbol and truth table for a two-input OR gate are illustrated in Figure 2.6. The output from the OR-gate is 0 only when both inputs are zero, and if any one input is 1, or if several inputs are 1, then the output is 1. This is not our everyday sense of OR, meaning one or the other but *not* both, so some care is needed if you are new to these ideas. Once again, the opposite type of gate is easier to make, and the NOR gate (Figure 2.7) is more usually found in IC form. The

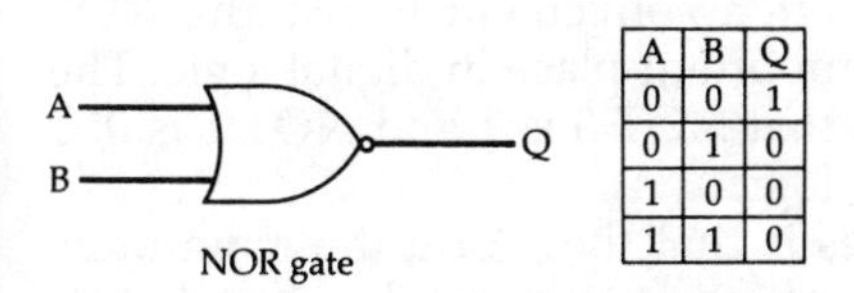

A	B	Q
0	0	1
0	1	0
1	0	0
1	1	0

Figure 2.7 The NOR gate, equivalent to the OR followed by an inverter. Like the NAND gate this can be used as a basic building block of al digital circuits.

more familiar type of one-or-other action is obtained from a different gate, the XOR (exclusive-OR) whose symbol and truth table is shown in Figure 2.8. This latter type of gate is extensively used in making gate circuits that carry out the arithmetic operation of adding two digital numbers.

There is another circuit which is usually included among the gates and referred to as the NOT gate, Figure 2.9. This is a digital inverter and though its action can be obtained from either the

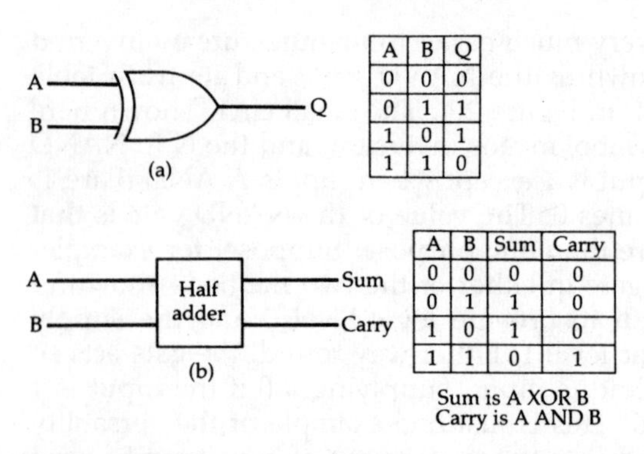

A	B	Q
0	0	0
0	1	1
1	0	1
1	1	0

A	B	Sum	Carry
0	0	0	0
0	1	1	0
1	0	1	0
1	1	0	1

Figure 2.8 (a) The exclusive-OR gate and its truth table. This action is closer to our idea of OR, because it excludes (hence the name) the case when both inputs are at 1. (b) How the action of simple addition of two bits can be carried out by an AND gate and an XOR gate.

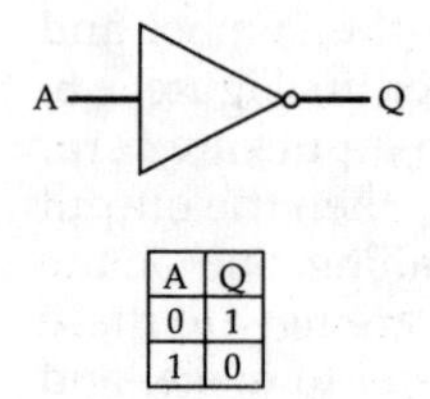

A	Q
0	1
1	0

Figure 2.9 The NOT gate symbol and truth table.

NAND or the NOR gate, it is often convenient to use the NOT symbol in circuits to match its important place in digital logic. The significance of the NOT gate is that NOT 0 is 1 and NOT 1 is 0, a vitally important part of digital logic.

The importance of these gates is that they form the main units of any digital system. Any type of logic action can be provided by using gates on digital signals, so that if, for example you wanted to obtain a 1 output when any two out of three inputs were at 1 it would be possible to design a suitable circuit using gates, and it would equally be possible to design the circuit from either the NAND or the NOR type of gates alone. Counting circuits can also be constructed from gates, so that the gate is to the digital circuit as the transistor is to the linear circuit. Knowledge of gates is not essential to the understanding of digital audio, but it's the simplest path into an understanding of the more advanced circuitry that is used in CD players, usually entirely in IC form. The logic actions that are needed for CD use include the track selection, track following, display of track availability, error correction, laser focus control, motor speed control, so that the logic circuits which are not involved directly in the audio reproduction system are nevertheless not exactly unimportant. Though you probably won't have to design them for yourself, or even service them, it's important to know what is involved, because the essence of gate logic is that any particular combination of inputs will produce an output which can be predicted from the truth table. Because of this, a logic circuit can be checked using DC steady voltages, and nothing elaborate is needed. It's very different when we get to sequential circuits.

Sequential circuits

Gate circuits are sometimes known as combinational circuits because what you get from the output of such a circuit depends on what combination of 0's and 1's happens to be present at the inputs. There is a very different type of circuit that can be built using gates, in which the output depends on the **sequence** of inputs rather than the combination. The simplest example is a counter IC, in which the state of the outputs will depend on how many pulses have arrived at the input. The classic type of sequential circuit is the flip-flop, though the digital gate version of this circuit is far removed from the simple type of direct coupled multivibrator that may be familiar to you. The most elementary form of flip-flop is known as the R-S flip-flop, and though it has few uses nowadays, it is incorporated into a lot of IC circuitry, and its action repays study. One type of RS flip-flop circuit using NAND gates, is illustrated in Figure 2.10.

Trying to understand what this circuit does looks at first like a monkey-puzzle. The key to understanding is to remember that the output of a two-input NAND-gate is zero only if both inputs are at 1, and will be 1 for all other inputs. In the circuit of Figure 2.10, if input R is 0, then, no matter what the signal level is at input S, the output of gate 1 must be 1. This means that if the other input at S is at level 1, then gate 2 has both inputs at 1 so that its output is also 0. If input S remains at level 1, changing input R to 1 will have no effect on the output at Q, because with the other input of gate 1 at 0, the output of gate 1 will still be 1. The action, then, is that a 1 at one input and a 0 at the other will set the output of the flip-flop one way or the other, and a 1 at each input 'locks' the output, leaving it in the state it had just before the change. This is

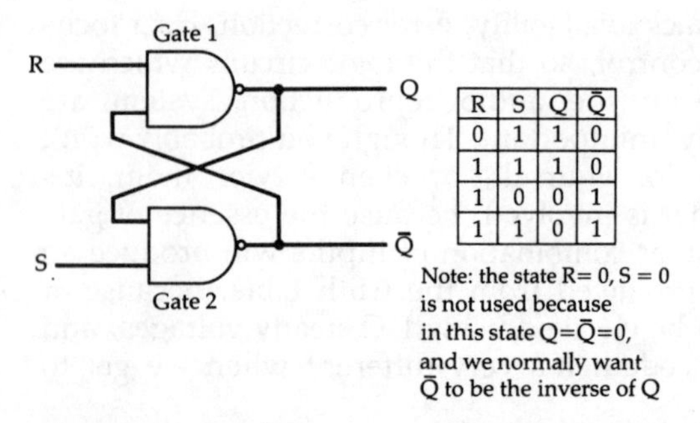

R	S	Q	$\bar{Q}$
0	1	1	0
1	1	1	0
1	0	0	1
1	1	0	1

Figure 2.10 The R-S flip-flop, the simplest form of sequential circuit. Unlike a normal (combinational) logic circuit, the flip-flop output depends on the sequence of inputs, not the combination. The R=1, S=1 state is a store state, maintaining the outputs at their previous value.

the essence of a sequential circuit – that the output depends on the sequence of inputs rather than the combination of inputs.

The basis of all sequential circuits is the flip-flop, and though simpler types like the R-S exist, the most important type of flip-flop is that known as the Master-Slave J-K flip-flop, abbreviated mercifully to JK. This is a *clocked* circuit, meaning that the action of the IC is carried out only when a pulse is applied to an input labelled 'clock'. This allows the actions of a number of such circuits to be perfectly synchronized, and avoids the kind of problems that can arise in some types of gate circuits when pulses arrive at different times. These problems are called 'race hazards', and their effect can be to cause erratic behaviour when a circuit is operated at high speeds. When clocking is used, the circuits are synchronous, meaning that each change takes place at the time of the clock pulse, and there should be no race hazards. Clock pulses, as the name implies, are generated at precisely equal intervals, usually from a quartz crystal controlled oscillator.

The symbol and a 'state table' for a JK flip-flop is illustrated in Figure 2.11. The table is a state table rather than a truth table, because the entries show that it is possible to have conditions in which identical inputs do not produce identical outputs, since the outputs depend on sequence of inputs rather than the voltages of the inputs at any particular time. The inputs are to the J and K terminals, and outputs are taken from the Q and bar-Q terminals. The bar-Q output is always the inverse of the Q output. The table shows the possible states of inputs before and after the clock pulse, showing how the voltages on the J and K pins will determine what

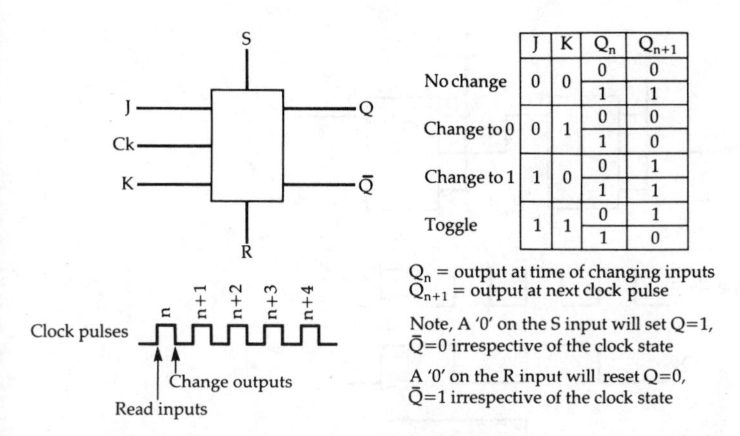

	J	K	Q_n	Q_{n+1}
No change	0	0	0	0
			1	1
Change to 0	0	1	0	0
			1	0
Change to 1	1	0	0	1
			1	1
Toggle	1	1	0	1
			1	0

Figure 2.11 The J-K flip-flop. The internal gate diagram has not been shown, only the conventional symbol. The leading edge of a clock pulse causes the J and K inputs to be read, and the output changes at the trailing edge of the clock pulse. The state table shows how the output will change from 0 or 1 for each possible combination of input at J and K. The J=1, K=1 input is called the toggling input and its effect is to reverse the output at each clock pulse.

the outputs will be after each clock pulse. Note that if both the J and K terminals are kept at level 1, the flip-flop will toggle so that the output changes over at each clock pulse, as is required for a simple binary counter.

Flip-flops are the basis of all counter circuits, because the toggling flip-flop is a single stage scale-of-two counter, giving a complete pulse at an output for each two pulses in at the clock terminal. By connecting another identical toggling flip-flop so that the output of the first flip-flop is used as the clock pulse of the second, a two-stage counter is created, so that the voltages at the Q outputs follow the binary count from 0 to 3, as Figure 2.12 shows. This principle can be extended to as many stages as is needed, and extended counters of this type can be used as timers, counting down a clock pulse which can be at a high frequency initially. The type of counter which uses toggling flip-flops in this way is called asynchronous, because the last flip-flop in a chain like this cannot be clocked until each other flip-flop in the chain has changed. For some purposes, this is acceptable, but for many other purposes it is essential to avoid these delays by using synchronous counter circuits in which the same input clock pulse is applied to all of the flip-flops in the chain, and correct counting is assured by connecting the J and K terminals through gates.

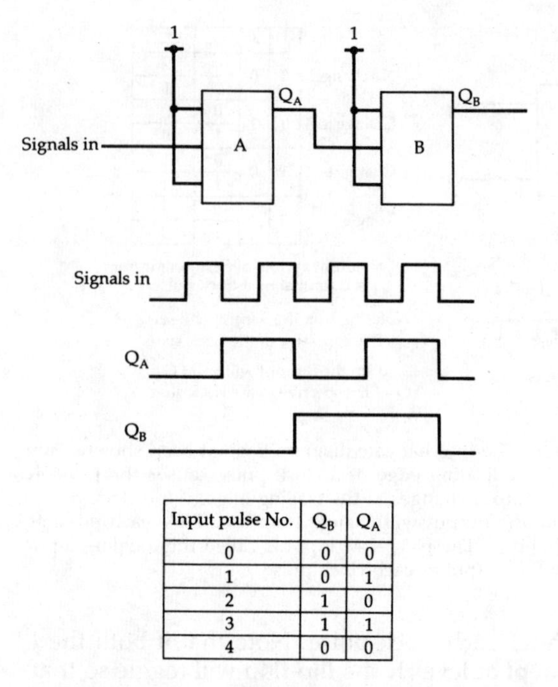

Input pulse No.	Q_B	Q_A
0	0	0
1	0	1
2	1	0
3	1	1
4	0	0

Figure 2.12 How two toggling J-K flip-flops connected together can form a counter. The output of the first flip-flop is taken to the clock terminal of the second, so that the second flip-flop is clocked at half the rate of the first. The signals diagram shows the effect of the division, which in terms of 1's and 0's is a binary count from 00 to 11. Note that flip-flop B represents the 2's figure (the most significant digit) and flip-flop A represents the least significant digit, the 1's figure.

Circuits of this type are beyond the scope of this book, and details can be found in the book whose title has already been quoted.

Registers

One of the main uses of flip-flops in digital circuitry is to form shift registers, usually in integrated form. A shift register, or more simply register, is an assembly of connected flip-flops and by altering the ways in which the flip-flops are connected we can create four different types of registers. These are known as SISO, SIPO, PISO and PIPO, with the letters S and P meaning serial and parallel respectively, and I and O meaning input and output. The actions of these registers are best explained by imagining a simple example of each type, containing four flip-flops with the output of each flip-flop at some level which will be either 0 or 1. The differences are concerned with how the inputs cause changes in the outputs, and the registers are always clocked, so that inputs will cause changes only at the time of a clock pulse, not when an input is supplied. The important feature of any type of register is that it will store digits, because each flip-flop will remain switched one way or the other until it is altered.

The first example in Figure 2.13 is a parallel-in, parallel out (PIPO) register. The flip-flops in such a register are connected only in the sense that the same clock pulse is used for each, but the action of each flip-flop is independent. At the time of the clock pulse, the input is copied to the output, and until the next clock pulse arrives the output remains unchanged no matter how the input changes. You can imagine this used as a sampling device,

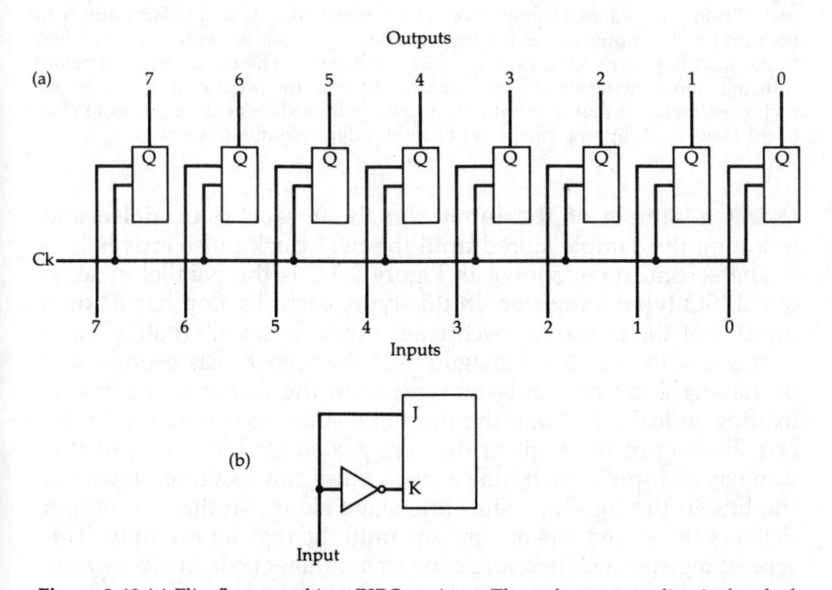

Figure 2.13 (a) Flip-flops used in a PIPO register. The only common line is the clock line, so that all of the flip-flops copy their inputs to their output together. Between clock pulses, the outputs remain unchanged regardless of changes at the inputs. The numbers 0 to 7 are the position numbers. corresponding to powers of two in a stored number and also to place significance, so that 0 means least significant, 2^0 (=1) and 7 means most significant, 2^7=128. (b) The input stage for a J-K flip-flop used in a PIPO register.

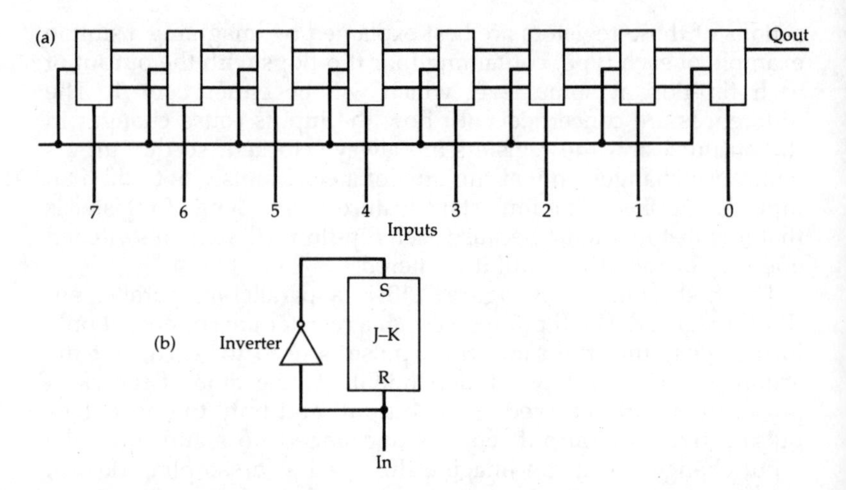

Figure 2.14 (a) The PISO register. The inputs are applied in parallel, one input to each flip-flop in order of significance. The connection of Q and Q-bar outputs to the next J and K inputs makes the register into a shift register, with each bit shifted to the next flip-flop on the right for each clock pulse. The clock pulses therefore shift the stored bits along the register and out from the serial output, one bit for each clock pulse so that this eight-bit register will need eight clock pulses to clear it. (b) How a J-K flip-flop can be set or reset independently of the clock.

taking a sample of the input signals at each clock pulse and retaining the sample stored until the next clock pulse arrives.

The second type, shown in Figure 2.14, is the parallel in serial out, PISO type of register. In this type, each flip flop has its own input, but the output of each is also used as an alternative input to the next in line. If we imagine that the register has been loaded by having a set of inputs present, then the action of a separate loading pulse is to make the flip-flops store each input, with the last flip-flop in the register delivering a single bit of output (the same as its input). Each time a clock pulse comes along, however, the bits in the flip-flops shift one stage along, so that the output delivers the stored bits one by one until the register is empty. This type of register will therefore convert a number code in binary form into a series of digital signals, and this is one crucial stage in converting a digital audio signal into a one-by-one serial form for recording on disc or tape.

The serial in, parallel out (SIPO) register of Figure 2.15 performs exactly the opposite action. A set of clock pulses will allow signals presented at the input to be accepted and shifted to the flip-flop

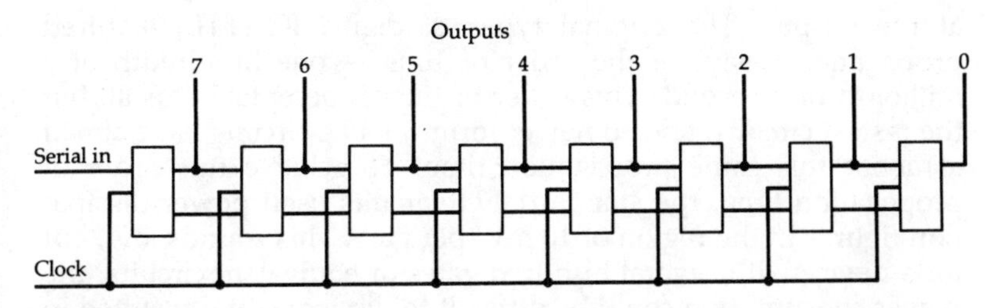

Figure 2.15 The SIPO register. This is also a shift register in which the bits are written in by clock pulses, eight in this example, and can from then onwards be read out from the parallel outputs.

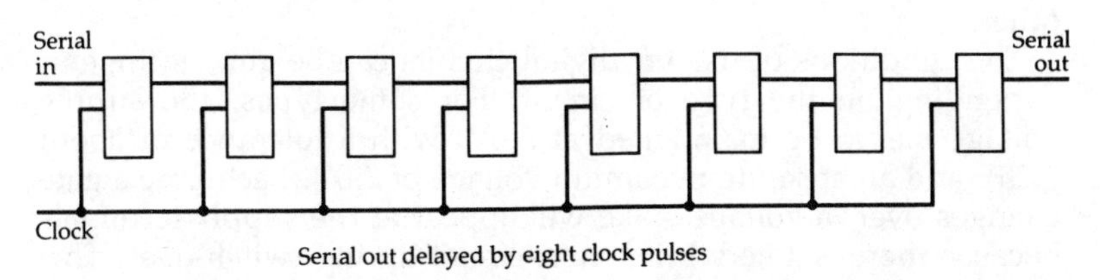

Figure 2.16 The SISO register which uses serial input and serial output. A set of clock pulses, eight in this example, will move a bit from the input to the output. This allows this type of register to be used as a memory in which the first bit in is also the first bit out (a first-in, first-out memory). The main use, however, is as a digital time delay, because the bits are delayed by the time of eight clock pulses. This technique is used in cross-interleaving (see later).

units in turn, so that these bits are available at the outputs. This converts a signal in serial form into one in parallel form, suitable for converting the signal read from a disc or tape into groups of bits that represent numbers.

Finally, Figure 2.16 shows the serial in serial out (SISO) register. This can be used as a time delay, because if, as in this example, the register consists of four flip-flops, then four clock pulses are needed in order to transfer an input signal bit to the output, with the flip-flops always storing four signals at any given time. The rate of the clock pulse signals will then determine how much delay is introduced, and this is the basis of many types of digital delay techniques used in audio processing, notably in the 'interleaving' process noted in Chapter 6.

Two important features of any IC for digital purposes are the propagation time and the power consumption per device. The propagation time is taken as an average figure for a gate, and means the time between changing the inputs and getting a change at the output. The original types of digital IC (TTL) featured propagation delays of the order of 10ns – one hundredth of a millionth of a second. This order of time is acceptable for all but the fastest circuits, and so newer forms of TTL circuits have aimed at about this same propagation time. To achieve this order of propagation time, the standard TTL circuits used power dissipation figures in the region of 10 mW per gate. This sounds low, but for a device with several hundred gates or equivalent circuits, the power consumption could be difficult to dissipate, and resulted in the chip running hot. The problem with the original type of design was that power dissipation and speed were connected – any reduction in one could be achieved only at the expense of the other.

The conditions of use for digital circuits can be quite stringent, depending on the type of circuit. For some types, the supply voltage has to be maintained at 5.0V, with a tolerance of about 0.25V and an absolute maximum voltage of 7.0V. Each time a gate changes over, a voltage spike will appear at the supply terminals because there is a very brief short-circuit during switch-over. This voltage spike can cause problems in other digital circuits, and so the 5.0V supply should always be stabilized. In addition, it is good practice to connect a capacitor of about 10nF between the supply and earth pins of each device, though if low-power devices are being used this can be relaxed to one capacitor for each five devices. The capacitors are particularly important in large circuits in which there may be long paths from the voltage regulator IC to the TTL circuits. Later types of IC have reduced the importance of the supply voltage and also reduced the amount of spike waveform, but the decoupling capacitors are still important.

The propagation times for digital circuits are affected by temperature, and by the types of load that can be connected. The worst type of load is a capacitive load with a pull-up resistor which is included to ensure that the voltage level rises to about supply-voltage level, as illustrated in Figure 2.17, because to reduce the

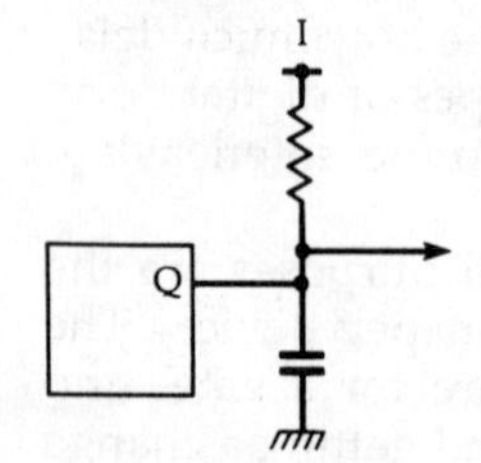

Figure 2.17 The worst type of load for a digital circuit, consisting of a pull-up resistor and a large stray capacitance.

voltage of this output to logic 0 means passing enough current through the resistor and also discharging the capacitor, even if the capacitance is only stray capacitance. The switching speed of the circuit will be affected by the amount of capacitance, typically by about 7ns for 80pF. This switching speed may also be affected by changes of temperature, but the effect of temperature is comparatively small.

The noise immunity of a digital circuit is an important feature, because it is a guide to what amplitude of interference may cause erratic operation of the circuits. Noise immunity is often quoted in terms of the size of interference pulse that will just cause a circuit to switch over. For many types of digital circuits, the noise immunity is determined by the small difference between 0 and 1 levels. If, for example, the maximum logic 0 voltage of 0.8V is used, and if the gate actually switches over at 1.4V (a fairly common figure), then an unwanted pulse of 0.6V will be enough to cause trouble. This makes some types of circuits, notably the older TTL types, very susceptible to interference on the supply lines, and is another good reason for using capacitors across the supply and earth terminals of each device.

CMOS ICs

The acronym CMOS means Complementary MOS, and complementary in this sense means that MOS transistors with both types of channels are used in pairs. A typical MOS circuit will use a P-channel FET in series with an N-channel FET, as illustrated in Figure 2.18. In this simplified basic circuit, the gate terminals of the FETs are connected together, so that an input signal, which will be a digital signal, will be applied to both gates. If we imagine

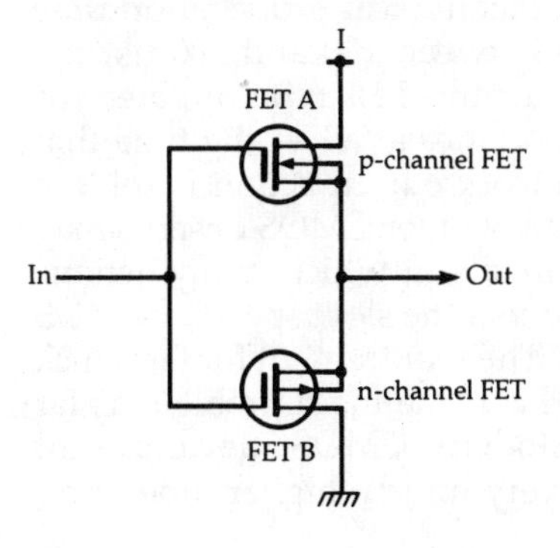

Figure 2.18 A typical MOS circuit element consisting of FETs of opposite type connected in series. For a digital input, one FET will be conducting and the other cut-off.

that the input signal is at logic 1, then the result will be to make FET A conduct fully, and FET B cut-off, with the result that the output is connected to the supply line at voltage level 1. If the input to the gates is at level 0, then the FET A is cut off, and the FET B is fully conducting, connecting the output to zero voltage, logic level 0. The basic circuit illustrated here is therefore one whose output is a replica of its input, neither inverting nor carrying out any other type of gating action. A circuit of this type can be used as an 'expander', allowing a larger number of loads to be driven from a single output.

The advantages of CMOS construction are that a wide range of supply voltages and logic voltages can be used. The standard and LS TTL circuits are designed to be operated with a +5V supply, and this voltage must be adhered to very closely. Most CMOS circuits will operate with voltages as low as 3V, some at 1.5V. They can also be operated with voltage levels of 12V or more, so that CMOS circuits can be used with low-voltage battery power equipment, and also can be applied with the voltage levels that are encountered in a car. The switching is also to 0 and 1 levels that are very close to the supply voltages. The construction of the output stages of TTL circuits makes it difficult to achieve a voltage for logic level 1 that is close to the +5V supply level, and which is more likely to be typically around 3.5V. This makes for a rather poor noise immunity for TTL, and because CMOS can achieve logic levels which are almost equal to the supply voltages, the noise immunity of CMOS can be very much better than that of standard or LS TTL. In addition, the power consumption of CMOS can be very much lower than that of TTL, of the order of a microwatt per gate.

The disadvantages of CMOS at one time were the slow operating speed of some early types, though this is not totally inevitable, and the need for electrostatic protection. The typical propagation time for the older type of CMOS gate is between 100 and 200 ns, ten times as much as the average time for the TTL type of gate. For many applications, this is of no importance at all – the time that is needed for a pocket calculator to work out an action is seldom found to be too long. Where the speed of older CMOS designs may be found unacceptable is in large circuits in which many actions have to be performed in sequence, so that the slow speed of CMOS will greatly increase the overall time. The requirement for fast clock rates can also be a problem – and this is the main problem as far as digital audio is concerned. Modern CMOS devices can, however, be manufactured with very much higher operating speeds, and we can also use devices in which the MOS channels are of the same polarity, such as PMOS and NMOS, which switch much faster than CMOS. The need for faster CMOS has been made necessary by the need for some memory chips in computers that can use very low power, and retain data when operated from a low-voltage battery.

In the early days of CMOS, electrostatic damage caused a lot of problems. The gates of MOS devices are so well insulated from the channels that even a tiny amount of electrostatic charge would not be discharged quickly, so permitting a build-up of voltage. At the same time, the very thin insulating layer of silicon oxide was easily punctured by excessive voltages, of the order of 20V or less in some cases. This means that handling a MOS device, or rubbing the pins against any insulating material or against textiles could cause total and irreversible breakdown of the gates. This looks as if it might be a considerable problem for anyone servicing digital equipment, but by the time that CMOS ICs were being used in any numbers, the circuits included discharge diodes at the gate terminals that ensured voltage limitation.

The important point to remember, though it is never mentioned by the manufacturers of 'anti-static' aids, is that only voltage between the gate and the channel can cause damage. It is possible that by walking across a nylon carpet, your hands can be at 15kV or more, but if this voltage is applied to both gate and channel, it does no harm. ICs that are correctly connected into a circuit are therefore immune from these static problems, because the circuits will have some resistance connected between gate and channel connections that will safely limit any conceivable electrostatic voltage. In other words, static should not cause any damage to a working circuit, whether switched on or off. The only possible case in which damage can occur in normal use is when one pin of a MOS IC is earthed and another pin touched, and even in this case, the built-in diodes should be able to cope.

In many years of handling MOS devices, I have never had one damaged by static, even on days when walking across the carpet and touching a radiator would result in a numbing shock. Figure 2.19 lists the standard precautions for handling MOS devices. These err, as always, on the safe side, and the precautions that are used in some assembly lines (earthed metal manacles for anyone handling MOS chips, for example) are not mentioned because they apply to specialised situations only. If ICs are never inserted or removed until equipment has been switched off and all voltages discharged, if ICs are kept in their protective packaging when not

1. Keep all MOS devices in the manufacturer's packing until needed. Try to avoid touching leads or pins.
2. Short circuit pins/leads together while installing, if possible.
3. Never allow a gate input to become open-circuit.
4. Use earthed soldering irons and earthed circuit boards while installing.
5. Ensure that circuit boards contain resistive paths between gates and other terminals (drain or source).
6. If devices have to be handled, ensure that surroundings are suitable – high humidity, no carpets of synthetic fibres, earthing on all metal surfaces. In severe conditions, earthed metal clamps may have to be fitted to the wrists of anyone handling MOS chips.

Figure 2.19 The precautions which are urged on users of CMOS and other MOS ICs.

in use, and if minor precautions are taken against static, like working with the hands slightly damp and with all circuit boards earthed, then problems due to static are most unlikely.

Figure 2.20 shows the general form of a CMOS gate. The input is always to gate terminals, and the output is taken from one drain and one source. Because the gate terminal of a MOS IC does not require current, the number of gate inputs that can be connected to a gate output (the **fanout** of the gate output) is limited only by the capacitance of the gate terminals. In other words, if the operating speed is very low, the fanout is almost unlimited, but for the higher operating speeds, the amount of current that has to be supplied in order to charge and discharge the gate-channel capacitance will be a limiting factor. The presence of the protective

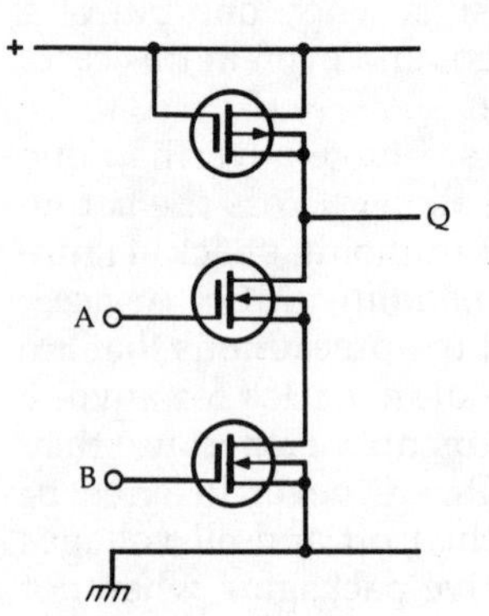

Figure 2.20 A CMOS NAND gate, illustrating that the inputs are always to the gates of the FETs, so that the input impedance is very high. The output is always from a source or drain, making this impedance fairly low and so allowing any output to drive a large number of inputs that are connected to it.

diodes also increases the amount of DC current leakage, though by a very small amount.

Though CMOS forms one important class of digital ICs, notably for its very low power consumption, many digital ICs use MOS FETs of one channel type only, such as NMOS and PMOS. These ICs offer low power consumption, though not as low as that of CMOS, along with fast operation. These types of FET ICs are used extensively in computing, both for microprocessors and for the supporting memory chips. These and CMOS ICs are also used in a variety of other applications for instrumentation and communications. Many of these ICs are of familiar digital types, such as counters, but with a greater degree of integration. Others are of a very specialised nature, like the speech chips that can be incorporated into equipment to permit spoken messages to be delivered, and the chips that form the signal-processing parts of compact disc and DAT systems. Many of the P-MOS and N-MOS ICs are intended for computing uses, and the popular microprocessor types are included in this heading. In this category of assorted circuits, we also have the 'bucket brigade', which is a type of register, a string of flip-flop devices in which data is shifted from one device to another in a chain at each input (clock) pulse. This device allows information to be delayed by a time that depends on the frequency of the clocking pulses, and the main uses are in obtaining delays for audio systems such as public address, music reverberation and special effects. The same principles, though in much more refined form, are used in the CCD (charge-coupled device) type of light-sensitive arrays that are replacing vidicon tubes in miniature TV cameras and camcorders.

Memory ICs

One very important class of MOS ICs for computing use is the memory type of IC. Memory in computing terms means the ability to retain logic 0 or 1 signals for as long as is needed, and the two classes of memory are volatile and non-volatile. Volatile memory is any type that requires a power supply to be switched on in order to retain data. All electronic memory is of this type, though some CMOS memory can be arranged to retain data almost indefinitely with the help of a miniature battery that is built into the circuit; sometimes a large-value electrolytic capacitor, typically 3F3, can be used for maintaining memory during power interruptions. The true non-volatile memory will retain data with all forms of voltage supply switched off. The most widely used form of non-volatile memory is the magnetic type, and the older type of large (mainframe) computers used tiny rings or cores of magnetic material, each one retaining one binary digit (or bit). For storage outside a computer, magnetic tapes and discs are used, and a more recent development is the optical type of disc as used in the CD system. The advantage of the CD type of disc is that a very much larger amount of data can be stored in a very compact form and with no risk that the data can be erased by careless exposure to magnetic fields, or by clumsy handling. It is no exaggeration to say that the impact of compact discs on computing is even greater than on audio, because it makes it possible to have huge amounts of data available for a modest-sized desktop computer. The likely impact of DAT (in whatever form it eventually becomes acceptable) is greater still, because unlike CD, DAT can be recorded as well as replayed and the system is ideally suited to backing up the data stored on a high-capacity hard disc in a computer.

Memory for short-term storage can use volatile memory, meaning devices like flip-flops that can store data only while a power supply is maintained. The types of volatile memory are classed as static or dynamic, and the two are radically different. Static memory (usually referred to as static RAM, because access to the memory can be random, to any part of the memory) makes use of a flip-flop for each bit that is to be stored. Since a flip-flop consists of two transistors, one of which is passing current while the other is off, each flip-flop in a memory of this type will pass current whether the bit that is stored is a 1 or a 0. Very early types of memory of this sort used the flipflops in a chain, clocked so that each clock pulse would shift each bit of data to the next flip-flop in line. This made for a very simple type of memory system consisting of just one flip-flop for each bit, but with limitations to access. For example, 1000 bits of data would be stored in a memory, assuming that it had at least 1000 flip-flops available, by applying the bits to an input along with a clock pulse for each input change. If the data was needed again, it had to be obtained at the output end of the chain, one bit for each clock pulse. A system like this is called first-in first-out, or FIFO memory. Each time the data is read out it must also be placed back into the flip-flops again if it is to be held for further use. In addition, all of the data has to be read – it is not possible to read the 56th bit without reading the first 55, for example, and if you need to keep the data then reading and writing must be carried out together.

Developments in IC techniques soon made it possible to connect inputs and outputs to any of the flip-flops in a chip. This is called 'random-access', and the name of random-access memory (RAM) was applied to such a chip to distinguish it from the earlier types. Random access demands the use of large numbers of gates on the chip, and also some method of selecting which flip-flop is to be accessed. The method that is used is to apply binary signals to address pins. The address of a bit is its reference number, the number of the flip-flop in which it is stored. When this number is selected by applying the binary signals for the number to a set of address pins on the chip, the correct gating is selected to make connection to that flip-flop, either for reading or for writing. The basic type of one-bit RAM chip therefore consists of one input pin, one output pin, a read/write pin, and a number of address pins. The use of 16 address pins, for example, allows 216 = 65536 different flip-flops to be accessed, so that this number of bits can be stored.

Any one flip-flop can be accessed by using its particular address number in binary form on the address pins. The use of flip-flops requires current in each flip-flop, and early types of static memory required a considerable amount of power. This led to the development of a different memory type, the dynamic memory, and though CMOS static memory can now be obtained with very low power consumption, the use of dynamic memory is now very well established, particularly in small computers. The principle of dynamic memory is storage of charge in a tiny capacitor constructed like the gate of a MOS transistor. Though the resistance between the contacts of such a capacitor can be very high, the small amount of charge that can be stored, along with the leakage through the selecting circuits of a memory chip limits the storage time to a few milliseconds only. This means that each tiny capacitor that stores a 1 must be recharged (or refreshed) at intervals of, usually, one millisecond. This is not so difficult as might be imagined, because the memory chips can be manufactured with the refreshing circuits built-in, and the master control of refreshing can be supplied either from a special controller chip, or in some cases by the microprocessor of the computer itself. The refresh has to be 'transparent' meaning that it should not interrupt the normal computing actions in any way.

The use of dynamic memory greatly reduces the amount of power that is needed for the memory of a small computer, because power is needed only to refresh the charge on the capacitors that are storing a logic 1 signal. The low power requirement makes it easier to build very large-capacity memory chips, so that the spectacular improvement in the memory that is available for small computers has been due largely to the availability of dynamic memory chips in ever increasing sizes. At the time of writing, the largest dynamic RAM in common use was of 1048576 bits. This is described as a 1 Mb chip, one megabit. In computing, 'K' is used to mean 1024 (which is 2^{10}) and M is 1024 × 1024 which gives the figure of 1048576, equal to 2^{20}.

The importance of temporary memory of this type is that it allows digital signals to be stored for processing. There is no need, for example, for the sampled signals to be recorded at exactly the time when they are sampled, and by retaining signals in memory for a time which is more than the time between sampling times, it is possible to compare successive signals. This is important when the sampled digital numbers are replayed because it allows one sample to be compared with another – in normal music, for example, one sample is not too vastly different from its neighbours, but a sample that is incorrect because of a fault on the disc for example will be very different and can be rejected. Such a sample can be replaced by a synthetic one which can be a number midway between the 'normal' samples, Figure 2.21. Using memory also allows samples that have been taken at the same time, such as samples of two channels of sound, to be transmitted or recorded at different times by storing one for later transmission, and later made coincident again by storing the one that had not been delayed the first time, Figure 2.22. It is even possible to store

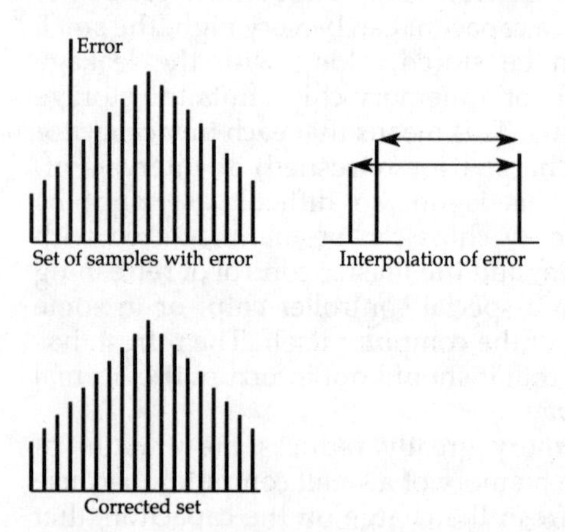

Figure 2.21 Using interpolation to replace a faulty sample with one that is mid-way between the amplitudes of the previous sample and the following sample.

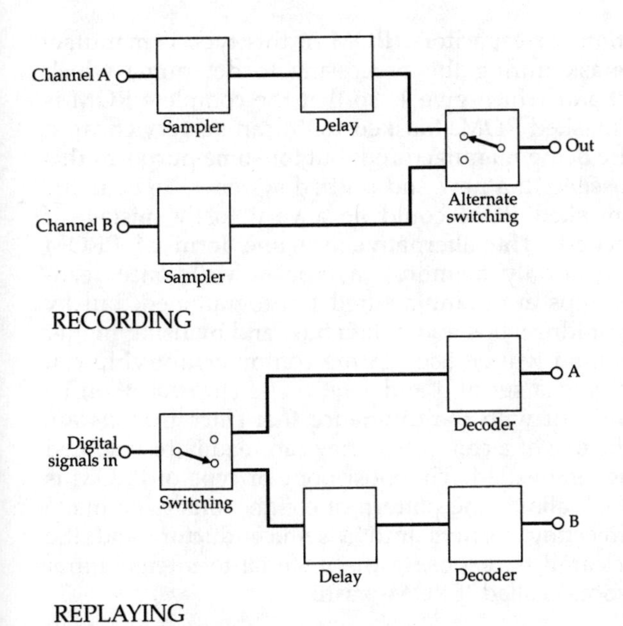

Figure 2.22 Using storage units to convert simultaneous waveforms into alternate samples and vice-versa.

groups of signals so that they need not be recorded in the order in which the signals arose; this is a very important part of the processing system of a compact disc, as we shall see, because it ensures that a fault will not affect more than a very insignificant portion of a piece of music.

In addition to this temporary memory which may be needed for short-time storing of signals we need a permanent memory system. We want, for example, to keep the signals that make the system work in some permanent form, because a digital audio system would be unacceptable if we had to start it working by feeding instructions into it from a disc, as we do with most types of computers. Memory of this type is referred to as ROM, read-only memory, and it can be of several forms. For the operating instructions of a CD player, the contents of the permanent memory will not be liable to any change, since there will be no significant changes in the CD system in the foreseeable future. The memory can therefore be programmed during manufacture, so that the gates that are operated by the address pin voltage will simply connect the data pin(s) to logic level 0 or 1 directly, not to the outputs of flip-flops or capacitors. ROM of this type is manufactured using a mask during the processing to determine which addresses give 0 and which give 1, so that the complete ROM is referred to as a masked ROM. Masked ROM can be very cheap if large numbers are being manufactured, but for some purposes this is not always possible. If a new and untried system is to be used, setting up for masked ROM could be a very costly mistake if changes are needed. The alternative is some form of PROM (programmable read-only memory), a type of write-once read-often chip. The chips are manufactured unprogrammed, but by connecting to an address bus and a data bus, and by using higher voltage supplies than will be used in the computer, the chip can be programmed with a set of 0 and 1 bits. The chip can then be used like a ROM, but with the difference that since the bits are established by the use of a computer, they can readily be changed if a design change is needed. The most popular type of PROM is the EPROM, which allows the pattern of connections to be made electrically by injecting carriers into a semiconductor, and the pattern can be cleared by exposing the material to intense ultraviolet light, a process called 'PROM-washing'.

Microprocessors

The microprocessor chip is the heart of any computer, and is now the universal component of a large number of industrial controllers. The chip consists of a large array of gates and flip-flops, and its unique feature is that internal connections can be made or broken by electrical programming. In other words, a set of binary signals at the data inputs of the microprocessor can be used to set the pattern of gates inside the chip and so determine the action that the chip will carry out with the next set of inputs. The important point to realise is that the microprocessor is a serial controller that carries out one action at a time. For example, if the microprocessor is to add two binary numbers, then it must go through a sequence of actions in which at least three inputs will be needed on the data lines. The first action is to place the ADD command in binary form on the data lines. Following this comes the first of the binary numbers to be added. The next item is the second number, and the output from the data lines will then consist of the sum of the numbers. This makes four distinct steps out of a simple addition, and illustrates why the use of a microprocessor can sometimes be too slow for some control actions. The speed at which the steps can be carried out is determined by a clock pulse rate, usually of several MHz. At the time of writing, speeds of 16 MHz and 20 MHz are fairly common for small computers, and 33MHz is being used on some machines.

Obtaining clock pulses

The final point about digital circuits concerns clock pulses. Most digital systems use clock pulses, and digital audio systems depend very heavily on the use of clock pulses, particularly to maintain the correct sampling rate as we shall see. This leads to the question of how these pulses are obtained in receiving equipment. The two main methods are to use a master crystal-controlled oscillator in each device, or to make the system 'self-clocking' by synchronizing an internal oscillator from sync pulses that are part of the received signal. Computer systems invariably use a master clock, because they must operate for long periods with no incoming data, and are just as likely to be recording data as replaying it. The two methods are not incompatible, and you can find the scheme of using sync pulses to correct a crystal oscillator, as is familiar from colour TV circuitry.

Both compact disc and DAT (digital audio tape) use self-clocking methods in which the master oscillator of the replay equipment is kept at the correct frequency and phase by a set of synchronizing pulses that are transmitted along with the audio data pulses. This is particularly important for tape, because the tape can stretch and contract to an extent which would make the use of a separate unsynchronized oscillator impossible. Even discs, however, need self-clocking because of the difficulty of ensuring that a disc is spun with perfect centring and with perfect speed regulation. By using self-clocking along with constant linear velocity (see Chapter 6) such problems become insignificant.

3 Analogue to digital conversions

Conversion and modulation

The conversion of analogue signals into digital form is the essential first step in any digital recording system. This step is part of the recording process rather than the replay process, but the methods that are used for this conversion affect very considerably how we can recover signals during replay, so that the principles need to be understood. Note, incidentally, that we have to distinguish between conversion and modulation in this context. Conversion means the processing of an analogue signal into a set of digital signals, and modulation means the change from the original digital signal into the type of digital signal that is used to convey the data and also how this is impressed on the medium, tape or disc. The two are, however, bound up with each other because many forms of conversion are also forms of modulation, and many are completely unsuitable for a digital recording system. The word 'modulation' is used in rather a confusing way in digital audio reference books, so that you need to be sure which meaning is intended.

Consider, for example, the type of modulation that is represented in Figure 3.1. In this system, an analogue signal is sampled at intervals to obtain a pulse whose amplitude is equal to the amplitude of the audio signal. This is little more than a digital version of amplitude modulation of a carrier, and it has about as much application to high-fidelity sound. The point is that in this case the conversion and the modulation system are one and the same – you can't separate them. The more modern methods are all based on a modulation system called **pulse code modulation**, in which the digital pulses represent the amplitude of the sample

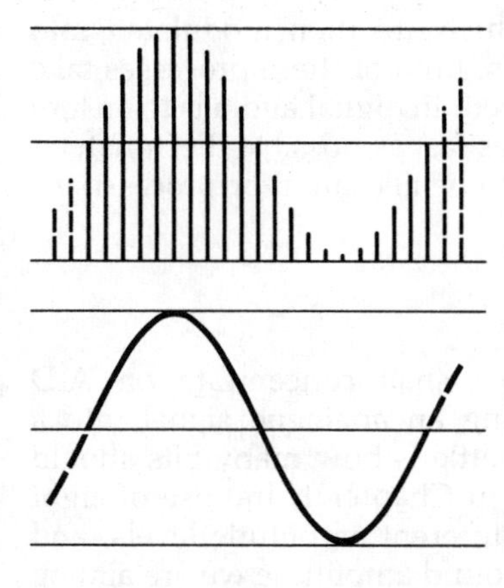

Figure 3.1 Illustrating pulse amplitude conversion and modulation. This system is of little use because it is as susceptible to interference as AM radio. Its only merit is simple conversion.

of analogue signal in a coded form, rather than by their amplitude, phase or repetition rate. This leads to the separation between the methods of converting the amplitude of the analogue signal into digital signals and the methods of organising these digital signals into a form that can be recorded, the other modulation part of the problem.

Leaving aside the matter of conversion for the moment, we have to try to see why modulation should be such a problem. The conversion will result in a number in digital form for each sampled piece of analogue signal, and this number will be proportional to the amplitude of the signal sample. Now from what we saw in Chapter 1, this digital number will be in binary form, using the digits 1 and 0, and it will be possible to have numbers that consist of mainly 1's and mainly 0's, such as 1000000000000000 and 0111111111111111. Now though these numbers cause no problems in the registers of a digital circuit, we have to think about how we could record them. A long string of 1's or 0's is to all intents and purposes a steady signal, like a DC level, and neither tape nor disc is a medium that can record DC. Whatever modulation method we choose must therefore be able to break up such patterns in such a way that changes from 0 to 1 or 1 to 0 take place often enough for the tape or disc system to respond correctly. In other words, what we are recording and replaying is the change from 0 to 1 or 1 to 0, and if these changes are too infrequent, there's a strong possibility that things will go wrong. The solution to this lies in the method of modulation of binary signals into a form of code that is less likely to cause such problems. This implies that signal amplitudes will be quantized into binary digits which are then modulated into another (longer) string of 1's and 0's. Both of these processes take time, and one of the problems inherent in digital audio is to ensure that such processes are carried out as fast as possible. Fortunately, storage of digits allows more time to be allocated for processing.

Conversion

From now on in this chapter we shall concentrate on A-D conversion, the process of changing an analogue signal into a digital number. The first thing to settle is how many bits should be used for a number. As we saw in Chapter 1, the use of eight bits permits us to distinguish 256 different amplitude levels, and this is insufficient for the range of sound amplitude we are aiming for. Putting this into concrete terms, we would want, in order to be considered of the highest audio standard, to be able to cope with a signal amplitude range of 90 db, which corresponds to a signal amplitude range of about 32000 to 1. Using just 256 steps of signal amplitude would make the size of each step about 123 units, too much of a change. This size of step is sometimes referred to as the quantum and the process as quantization. In these terms, then, 256 steps is too coarse a quantization for good quality of sound.

If we move to the use of 16 bits, allowing 32767 steps, we can see that this allows a number of steps that is well matched to the amplitude range of 90 db or 32000:1. This is the quantization that has been used in the CD system therefore, and also in digital tape systems. We would use the smaller number if we could, because working with sixteen bits takes longer when the bits are recorded one by one, and particularly when we use the modulation method that will be described in Chapter 6. It is this choice of coding along with the choice of modulation system that makes the operating frequency of a CD system so very high, equivalent to video frequency signals. Smaller quantization numbers have been used for TV stereo transmissions by compressing the digital signals, and these techniques could be usefully employed on discs or tape if it were possible to convert existing equipment, or if a new format were introduced (as has been done in the audio circuits of the Video-8 recorders).

There's more to the choice of number of bits than meets the eye, however. Ideally, the amplitude of a signal at each sample would be proportional to a number in the 16-bit range. Inevitably, this will not be so, and the difference between the actual amplitude and the amplitude that we can encode as a sixteen bit number represents an error, the quantization noise. We don't have to resort to any elaborate mathematical proofs to see that the greater the number of bits we use to encode an amplitude, the lower the quantization noise will be.

What is less predictable without theory, however, is the effect on low-amplitude signals. For low-amplitude signals, the amount of quantization noise is virtually proportional to the amplitude of signal, so that to the ear it sounds less like noise, and more like distortion. The greater the number of digital bits you use for encoding, the worse this effect gets unless you do something about it. The cure, by a strange paradox, is to add noise! Adding white noise – noise whose amplitude range is fairly constant over a large frequency range – to very low amplitude signals helps to break the connection between the quantization noise and the signal amplitude and so greatly reduces the effects that sound so like distortion. This added noise is called 'dither', and is another very important part of the conversion process. The noise level is very low, corresponding to a one-digit number.

Sampling rate

All A-D conversion, however, starts with sampling. Sampling means that the amplitude of the audio analogue signal is measured and stored in a short interval of time, and if sampling is to be used as a part of the process of converting from analogue to digital signals then it has to be repeated at regular intervals. The principles of the process are illustrated in Figure 3.2, from which you can see that if you intend to convert an analogue waveform into digital form with any degree of fidelity, then you need to take a large number of samples in the course of one cycle.

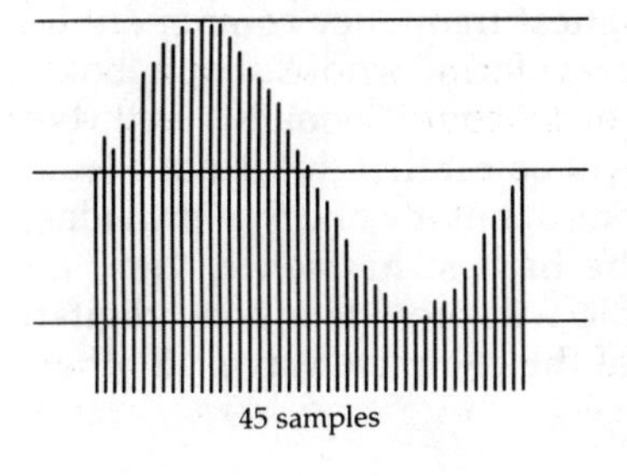

Figure 3.2 A waveform sampled 45 times, indicated by the dotted vertical lines. At this density of sampling, very fine detail of the waveform can be captured.

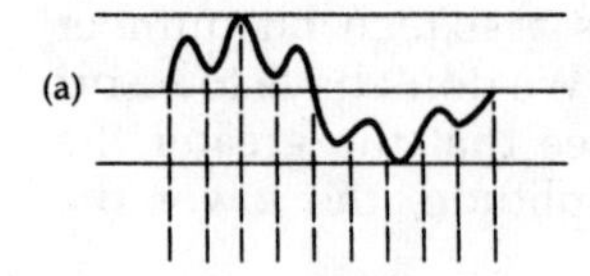

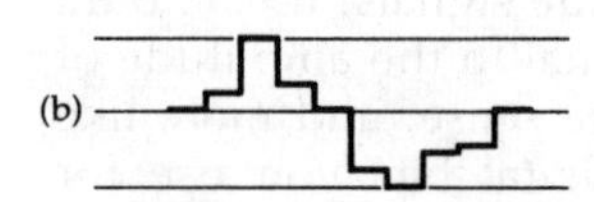

Figure 3.3 Inadequate sampling of the original wave (a) causes the replayed wave (b) to be block-shaped, and a poor replica. This is less significant if the amplitude of such portions of a wave are very small, which in practice they always are.

If you take too few samples, then the digital version of the signal (Figure 3.3) will look quite unlike the analogue version. On the other hand, if you take too many samples per cycle, you are working with a lot of redundant information and wasting disc space. Whatever sampling rate you use must be a reasonable compromise between efficiency and fidelity, and as it happens, the theory of sampling is by no means new so that we don't have to look far for our answers.

The idea of digital communications has been around for some time now, and in 1948, C. E. Shannon had published his paper 'A mathematical theory of communications' that the whole of digital audio is based around. The essence of Shannon's work is that if your sampling rate is twice the highest frequency component in an audio signal, then the balance between fidelity and excessive bandwidth is correctly struck. Note that this pivots around the highest frequency component. If you were working with a 1kHz sinewave, you would not achieve digital conversion of hi-fi standard by using a sampling rate of 2kHz, unless your reverse conversion was arranged always to provide a sinewave output. What Shannon's theory is about is non-sine waves, the sort of waves we are working with in audio, which can be analysed into a fundamental frequency and a set of harmonics.

What this boils down to is that if we look at a typical audio waveshape (Figure 3.4), then the highest frequency component is responsible for a small part of the waveform, whose shape looks nothing like a sinewave and which could equally well be represented by a sawtooth. Purists will object that this is not a true representation of the effect of harmonics, but it's one that provides a good mental picture. Sampling the highest **harmonic**, then, at twice the frequency of that harmonic will provide a good digital representation of the overall shape of the complete wave, all other things being equal.

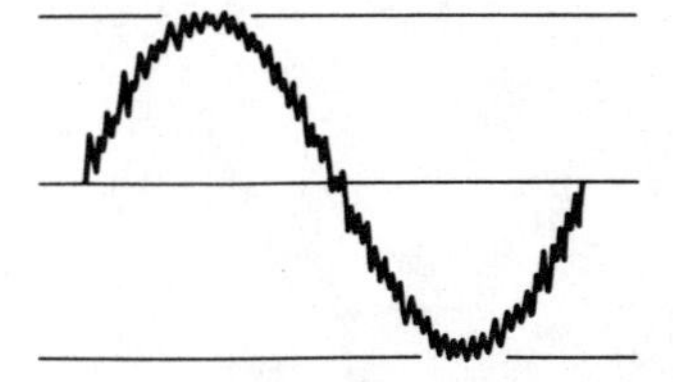

Figure 3.4 A representative audio waveform. Sampling will give a satisfactory version of this because the approximations affect only the smallest amplitudes of high-frequency signals, portions which many analogue systems filter out in any case.

Sampling under these conditions provides a set of pulses whose amplitude is proportional to the amplitude of the wave at each sampled point. If we use a spectrum analyzer to see what frequencies are present in such sampled signals we find something like the illustration of Figure 3.5. This consists of the range of frequencies that were present in the analogue signal (the fundamentals) plus a set of harmonics centred around the sampling frequency and its harmonics. This is not a problem when the pulse amplitudes are being converted into digital form, but when the pulses are recovered, a filter is needed to separate out the wanted part, which is the lowest range of frequencies, the frequencies of the original signal. The nature of this problem and its solution will be dealt with in Chapter 4.

The existence of these harmonics makes it important to ensure that the sampling rate is high enough. Suppose, for example, that we wanted to work with audio with an 18 kHz bandwidth and we picked a sampling rate of 30 kHz. The harmonics around this 30 kHz sampling frequency would extend down to 30 − 18 = 12 kHz

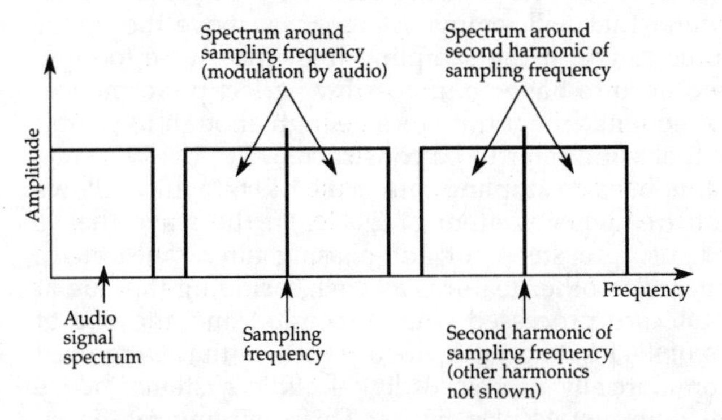

Figure 3.5 The spectrum (plot of amplitude against frequency) for a sampled waveform shows the frequencies of the original audio signal, plus the sampling frequency and sidebands (as wide as the audio signal) and a set of harmonics of the sampling frequency and corresponding sidebands.

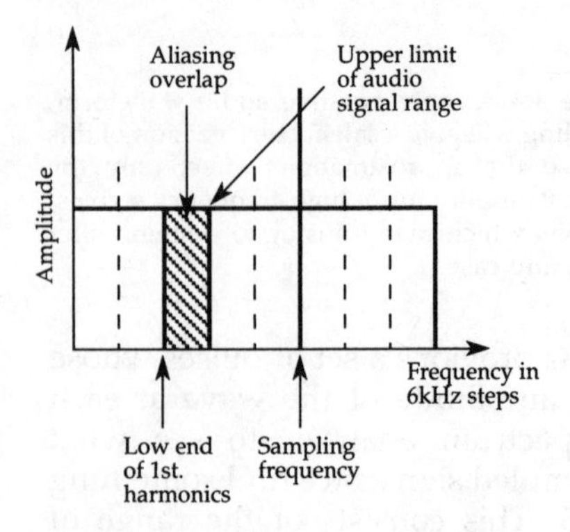

Figure 3.6 Aliasing – the result of using a sampling frequency that is too low. The upper frequencies of the audio signal overlap the lower frequencies of the lower sideband of the sampling frequency.

and up to 30 + 18 = 48 kHz, but it's the lower sideband of this set that is worrying, because it overlaps the 18 kHz of the original sound, Figure 3.6. This is an effect called 'aliasing', meaning that over a range of frequencies in the original range there will be a set of 'aliases' from the lower sidebands of the sampling frequency.

Even if the sampling frequency is made twice the highest audio frequency, difficulties still arise because there may be harmonics in the audio signal that extend to higher than half of the sampling frequency. This is dealt with by using an *anti-aliasing* filter which is a steep-cut filter that will remove frequencies above the upper limit of the audio range. If the sampling frequency were too low this filter would need to have an impossibly perfect performance. The sampling frequency must therefore be high enough to permit an effective anti-aliasing filter to be constructed.

The CD system uses a sampling rate of 44.1 kHz, which allows for a maximum frequency content of 20kHz in the wave that is being sampled, using a steep-cut anti-aliasing filter. The rate is chosen with regard to other features as well, including the rate at which signal bits are processed (the clock rate) and the use of sampling rates of 44.1 kHz on the video recorders that were used for the first commercially produced digital audio systems, hence the 44.1 rather than just 44 kHz figure. This sampling rate is, in fact, on the generous side, reflecting the determination of the original designers of the CD system to improve on rather than just simply match the best of conventional recording methods.

The sampling process

The process of sampling involves the use of a sample-and-hold circuit. As the name suggests, this is a circuit in which the amplitude of a waveform is sampled and held in memory while the amplitude size can be converted into digital form. The outline of a sample-and-hold circuit is shown in Figure 3.7, with a

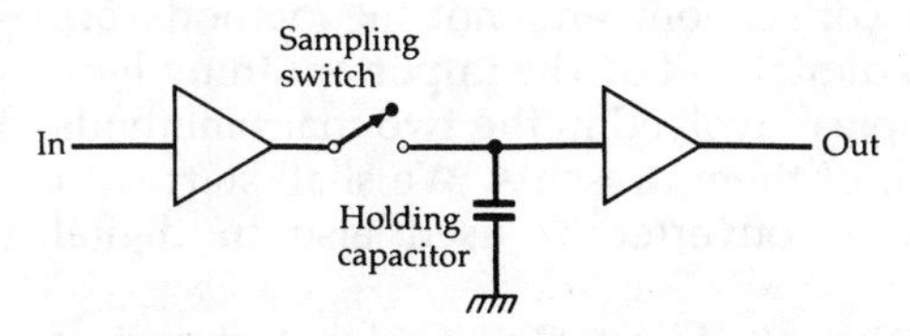

Figure 3.7 The sample and hold action represented by a switch and capacitor. The switch will be an analogue switch, like a FET bridge, but the capacitor can be used for storage provided that the input impedance of the second buffer is high.

capacitor representing the holding part of the process. While the switch is closed, the voltage across the capacitor is the audio voltage, maintained by the buffer amplifier which has a low output impedance. No conversion to digital form takes place in this interval. When the switch is opened, the amplitude of the audio signal at the instant of opening is the voltage across the capacitor, and this amplitude controls the output of the second buffer stage. This in turn, is the signal that will be converted to digital form.

The instant of sampling can be very short, then, but the time that is available for conversion to digital form is the time between sampling pulses. The 'switch' that is shown can be a semiconductor switch, and the capacitor can be a semiconductor memory, though for a fast sampling rate like 44.1 kHz, a capacitor is perfectly acceptable when its only loading is the input impedance of a MOS buffer stage. At a sampling rate of 44.1 kHz, the time available between sampling intervals is around 22 μs, quite long for conversion to digital form by the standards of modern equipment.

Conversion

The effect of sampling is only to quantize the signal. The signal is still an analogue signal in which the variation of amplitude with time carries the information of the signal, and the change that has come about as a result of sampling it is the substitution of a set of pulses for the original smoothly varying (if any audio signal can be called smoothly-varying) signal. The signal is now an amplitude-modulated set of pulses at the sampling frequency – but this is *not* a digital signal.

The actual conversion from analogue to digital form is the crucial part of the whole encoding process. There is more than one method of achieving this conversion, and not all methods are equally easily applicable to digital audio. The important thing here is to understand the principles involved in the two main methods and the problems that each of them presents. We shall start with the integrator type of A-D converter as used also in digital voltmeters.

The principle is simple enough, Figure 3.8. The central part of the circuit is a comparator, which has two inputs and one output. While one input remains below the level of the other, the output remains at one logic level, but the logic level at the output switches

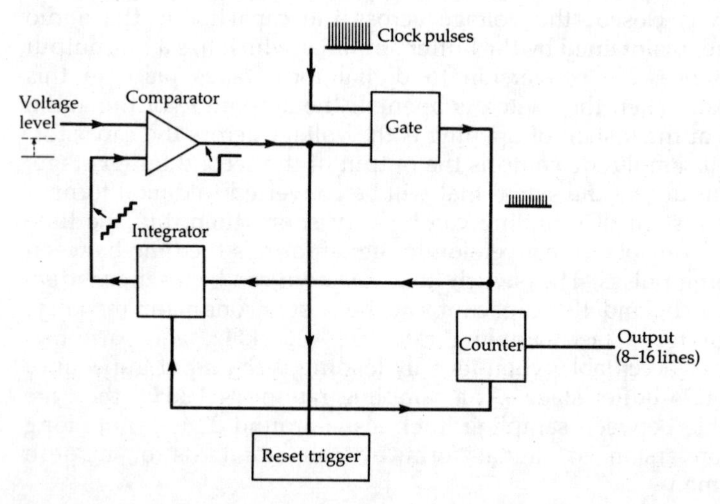

Figure 3.8 Principle of A-D conversion using an integrator and comparator. The clock pulses are passed through the gate and are counted and integrated. When the level of the integrated signal equals or exceeds the input signal amplitude the comparator shuts off the gate. The output from the counter (on a set of binary coded lines) can be read, and then the reset trigger will reset the counter, integrator and gate. The amplitude of the input is assumed to remain constant during the time needed for the count, so that a fast clock rate is needed.

over when the input levels are reversed. The change at the inputs that is needed to achieve this can be very small, a matter of millivolts, so that the action as far as we are concerned is that the output switches over when the input levels are equal.

If one input is the signal we are converting, this signal will be held at a steady level (by the sample and hold circuit) during the time that is needed for the conversion. A series of clock pulses are then applied both to a counter and to an integrating circuit. The output of the integrating circuit is a series of equal steps of voltage, rising by one step at each clock pulse, and this waveform is applied to the other input of the comparator. When the two inputs are at the same level (or the step waveform input just exceeds the sample input), the comparator switches over and this switch-over action can be used to interrupt the clock pulses, leaving the counter storing the number of pulses that arrived.

Now suppose that the steps of voltage were 1 mV, and that the sample voltage was 3.145 volts. With the output of the integrator rising by 1 mV in each step, 3145 steps would be needed to achieve equality and so stop the count, and the count would be of the number 3145, a digital number that represents the amplitude of the voltage at the sampled input. The switch-over of the comparator can be used to store this number into a register as well as stopping the clock pulses. A master clock pulse can then reset the integrator output to zero and terminate the conversion ready to start again when a new sample has been taken.

These numbers are, of course, for illustrative purposes only, but they demonstrate well how a signal level can be converted into a digital number by this method. Note that even with this comparatively crude example, the number of steps is well in excess of the 256 that we could cope with using an 8-bit digital number. Note also that the time that is needed to achieve the conversion depends on how fast a count rate is used — this will obviously need to be much faster than the sampling rate, since the conversion of each pulse should be completed by the time the next pulse is sampled.

The quality of conversion by this method depends very heavily on how well the integrator performs. An integrator is a form of digital to analogue (D-A) converter, so that we have the paradox here of using a D-A converter as an essential part of the A-D conversion process — rather like the problem of the chicken and the egg. The very simple forms of integrator, like charging a capacitor through a resistor, are unsuitable because their linearity is not good enough. The height of each step must be equal, and in a capacitor charging system, the height of each step is less than that of the step previous to it. The integrator is therefore very often a full-blown D-A converter in integrated form, using the principles that we shall be looking at in the following chapter.

Now for the headaches. Assuming we have mastered the problems of achieving perfect integration with a D-A converter, can the circuit cope with the speed at which it will have to be operated? This speed depends on the time that is available between samples and the number of steps in the conversion. If we have 20μs available to deal with a maximum level of 65536 steps, then the clock rate for the step pulses must be:

$$\frac{65536}{20 \times 10^6}$$

– which gives a frequency of 3.8 GHz (**not** Mhz), well above the limits of conventional digital equipment, and most other consumer equipment apart from satellite down-converters. This makes the simple form of integrator conversion impossible for the sampling rate and number of bits that we need to use for digital audio work, and we have to find ways around the problem. One method, if we want to stay with the integrator type of converter, is to split the action between two converters, each working with part of the voltage. The idea here is that one counter works in the range 0 to 255, and the other in units that are 256 times the steps of the first. The voltage is therefore measured as two eight-bit numbers, each of which requires only 255 steps, so that the counting rate can be considerably lower. Since the counters work in succession, with the smaller range of counter operating only after the coarser step counter has finished (Figure 3.9), the total number of steps is $2 \times 255 = 510$, and the step rate is now, still assuming a 20μs time interval:

$$\frac{510}{20 \times 10^6}$$

– a rate of 25.5 MHz. This may look a terrifying frequency if you are used to audio frequencies or indeed to the comparatively slow clock rates of the personal type of computer, but it is well within the range of modern digital circuitry in IC form.

Another form of A-D converter is known as the successive approximation type. The outline of the method is shown in Figure 3.10, consisting of a serial input parallel output (SIPO) register, a set of latches (otherwise known as a PIPO register), a D-A converter and a comparator whose output is used to operate the

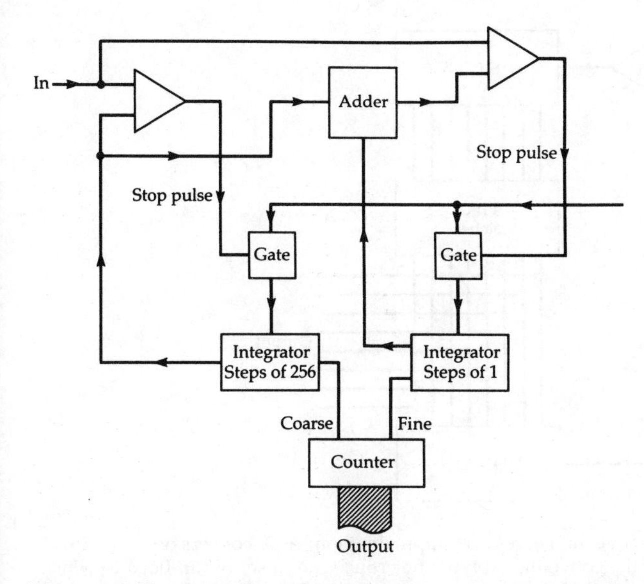

Figure 3.9 The principle of the two-step integrator system, details omitted. The coarse step integrator produces an output whose level is the nearest whole number of steps that is just less than the input, and this number is counted to give the higher set of binary signals. The difference between this output and the input is used to operate the fine step integrator to give the lower set of bits. The coarse step comparator must be biased so that its output will switch at a voltage lower than the actual input, and its output is also used to start the fine-step integrator.

register. This is a type of circuit whose action can be very obscure unless you go through it in easy steps, so we shall imagine what happens in the course of a sampling period.

Imagine first of all that the output from the D-A converter is zero. and therefore less than the input signal at the time of the sample. The result of these two inputs to the comparator is to make the output of the comparator high, and the first clock pulse arriving at the SIPO register will switch on the first flip-flop of the PIPO register, making its output high to match the high input from the comparator. This first PIPO output is connected to the highest bit input of the D-A converter which for a 16-bit register corresponds to 32767 steps of amplitude.

Now what happens next depends on whether the input signal level is more than or less than the level of output from the D-A converter for this input number. If the input signal is less than this, the output of the comparator changes to zero, and this will in turn

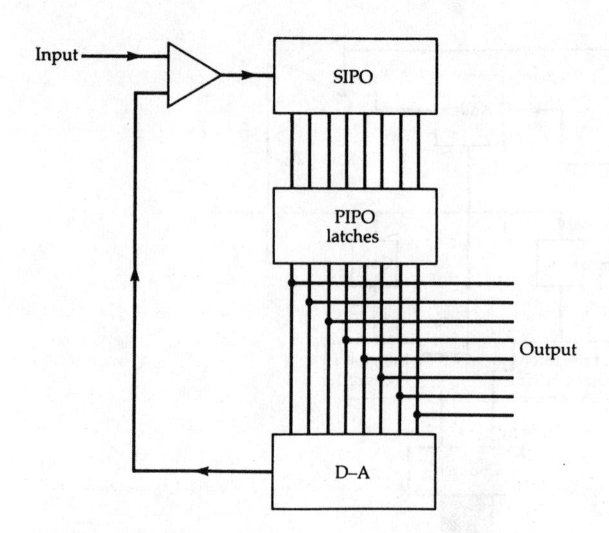

Figure 3.10 Principle of successive approximation A-D conversion. The PIPO register is not of the conventional type, but consists of a set of flip-flops to which the clock pulse is applied in sequence, not together. This allows the outputs to be changed one after the other without affecting previous outputs.

make the output of the SIPO register zero and the output of the PIPO register zero for this bit. If the input signal is greater than the output from the D-A circuit, then the 1 bit remains in the first place of the register. The clock pulse will then pass this pulse down to the next PIPO input – this does not, however, affect the first PIPO input which will remain set at the level it had attained.

Another comparison is now made, this time between the input signal and the output of the D-A converter with another input bit. The D-A output will either be greater than or less than the input signal level, and as a result, the second bit in the PIPO register will be set to 1 or reset to 0. This second bit represents 16384 steps if its level is 1, zero if its level is 0. The process is repeated for all sixteen stages in the register until the digital number that is connected from the PIPO outputs to the D-A inputs makes the output of the D-A circuit equal to the level of the input signal.

Using this method, only **16** comparisons have to be made in the sampling period, giving a maximum time of 1.25μs for each operation. This time, however, includes the time needed for shifting bits along the registers, and it requires a fast performance from the D-A converter, faster than is easily obtained from most designs. Speed is the problem with most digital circuits, which is why there is a constant effort made to improve methods of manufacturing ICs, and even to alternative semiconductor materials (like gallium arsenide) that could allow faster operation.

Negative numbers

So far, we have been assuming that the numbers we work with are in the range of 0 to 65535, the range of a number that can be represented by sixteen bits. This makes no allowance for the use of negative numbers, but the amplitudes that we are dealing with will be alternately positive and negative. The binary number system makes no direct provision for the use of a negative sign, however, and any method that we use for converting denary numbers into binary must use one of the binary bits itself to represent a negative sign.

The convention that is adopted in computing is the very common one of 2's complement, in which the highest order bit is used as a sign bit, 0 for positive and 1 for negative. The principle is illustrated in Figure 3.11 for a four-bit number, showing that numbers from 0000 to 0111 (0 to +7) will be positive and numbers from 1000 to 1111 (normally 8 to 15) will be negative. The way that the numbers progress, however, is slightly unexpected, because in this simple example, the value changes from +7 to −7 with the change from 0111 to 1000. The value of a negative number is, in denary terms, the value of the number (ignoring sign) with 16 subtracted.

All this arises from the method that is used to form a negative number in this conventional binary system. The number must be in the range that does not use the highest-order bit, and the conversion to negative form is done by inverting the digits (changing 1 to 0 and 0 to 1) and then adding 1 to the lowest place. For a sixteen-bit number this implies that the range of positive numbers is from 0 to 32767 and the range of negative numbers is from −1 to −32768. The same absolute range of 0 to 65535 is still being used, but split into positive and negative sections. The convention about how we regard a number does not affect the methods that are used for A-D conversion, and all we need to do is to ensure that the polarity of the signals will be such that negative voltage amplitudes generate numbers greater than 1000000000000000 in the A-D converter.

This method of representing a negative number, used almost

Four bit range is −8 to +7
Conversion of positive numbers as before
for negative numbers

1 Convert as if positive	eg −6→ +6 = 0110
2 Invert each digit	eg 0110→ 1001
3 Add 1	eg 1001 + 1 = 1010
4 This is negative binary number	

For 4 bit number

0	0000
1	0001
2	0010
3	0011
4	0100
5	0101
6	0110
7	0111
−8	1000
−7	1001
−6	1010
−5	1011
−4	1100
−3	1101
−2	1110
−1	1111

Figure 3.11 The 2's complement method of coding a negative number in 8–4–2–1 code. The most significant bit (left hand side) is used as a sign bit, rather than just as a number bit. The conversion consists of converting regardless of sign, and then for a negative number inverting and adding one. The sequence of codes shows how the value changes violently with the change from 0111 to 1000 in this example.

universally in digital systems, brings its own headaches. A low-level signal, for example, will be changing from a small positive number such as 0000000000000001 to a small negative number such as 1000000000000001 in which just one bit, the most significant bit, has changed. The effect of this causes problems in the conversion from digital to analogue, as we shall see later. The other problem lies in coding. All of these numbers are likely to contain long runs or either 0's or 1's, which will cause problems with any kind of recording system. This means that the number have to be converted, using a coding system which:

1. Does not allow long 'DC' intervals of identical bits.
2. Does not allow transitions to occur close to each other

This second condition needs more explanation. If any coding system allows a sequence such as 1010101. . . . to occur, there are bound to be problems with both recording and replaying because of the difficulty of correctly tracking such rapid changes – remember the rate at which these pulses have to be read and written. Given these problems, then, the conversion to binary number is just one step in the record/replay system, and the binary numbers are neither recorded nor replayed directly. As we shall see, digital audio systems usually ignore negative numbers by using a method rather like the use of bias, taking a number midway between zero and maximum as the real zero, sometimes referred to as digital silence. Using this system, negative numbers have a most significant bit of zero and positive numbers have a most significant bit equal to 1.

Error checking

Inevitably in any recording system there will be errors. On tape or disc there will be dropout errors where because of a tiny fault in the recording material some signals cannot be recorded. On magnetic tape, these dropouts will be caused by missing or poor-quality magnetic coating, and though the incidence of dropouts can be made small it can never be eliminated; a similar situation arises in any method of disc recording. Analogue recording suffers from dropouts, but if the dropouts are of very short duration they are virtually inaudible. For digital recording, however, the duration of a dropout could result in hundreds of digits being missed, making the problem more serious.

Another problem is jitter, analogous to wow and flutter in analogue recording. Jitter means that the imperfect control of speed of the recording medium will cause alteration in the frequency with which signals are replayed, and if the system depends, as all digital systems do, on the maintenance of a set clock frequency, some method will have to be found of keeping this clock frequency in step with any jitter.

Another form of error is more subtle and specific to digital recording methods – it is called intersymbol interference. However we record a pulse, it is most unlikely that the recorded form will be as narrow as the pulse was, whether the recorded form is a magnetic signal or a mark on a disc. This brings the danger that adjacent pulses can interfere with other, and is another reason for the requirement that we need to avoid sequences such as the 101010. . . type mentioned earlier.

Other forms of error are from the familiar old enemy of all recording, noise, whether caused by the analogue parts of the processing, the digital portions, or by the effects of editing master tapes or even blunders in recording or processing at the studios.

One of the problems of any digital number system, irrespective of the system used, is that an error that causes a 1 to change to 0 or 0 to 1 can make a very marked difference to a number. If the error takes place so as to affect a low order digit, this is not particularly serious – for example if the number 0110111001011011 (denary 28251) changes to 0110111001011010 (denary 28250), then the change is only one part in some twenty-eight thousand, negligible. If, however, the change is to a higher order bit, perhaps in the example above to 0010111001011011 (which is 11867 in denary) then the change is very large and the error is certainly not negligible. It is even more important if the change is to the highest order bit, changing the number from positive to negative or vice versa as well as changing its value.

Any digital system must therefore make some provision for error checking, and for digital audio checking alone is not enough. There must be some type of provision made to **correct** any error, because the nature of all recording materials, whether tape or disc, must be that bits of digital numbers will inevitably go missing. In using tape we always have the problem of dropouts and dropins which will affect the recording and reading of digital numbers, and on discs we have the problems of scratches and other imperfections of the surface. A scratched disc may make a noticeable noise in an analogue replay, but it can make a very much greater impact on a stored number, though this may, as we shall see, not be quite such a significant part of the waveform.

At this stage, we shall not go deeply into error detection and correction, because this is one of the most difficult aspects of digital audio and one that most users never come across because the whole process is controlled as part of the signal processing system. We need, however, to know the principles that are involved. These boil down to digit counting, redundancy and level maintenance.

Digit counting is a method of checking for the presence of one incorrect digit, and the simplest system is the parity system as used on serial transmission links between computers and printers. We noted this in Chapter 1, but a concrete example will make the advantages and disadvantages more clear. Suppose, for the sake of simplicity, that we are dealing with numbers of three digits only, in the range 000 to 111. Now such numbers can contain either an even or an odd number of 1's, like 001 and 011, and the presence of an odd or even number of 1's can be detected very simply by a set of XOR gates. Achieving parity means adding an extra bit so that all numbers contain either an even or an odd number of 1's.

In the even parity system, each number is made up with an extra digit (placed in the highest position) which will make the number of 1's always even. This the number sequence of our example, 000, 001, 011, 100, 101, 110, 111 would become 0000, 1001, 0011, 1100, 1101, 0110, 1111 respectively. If we chose an odd parity system, each number would need to contain an odd number of 1's, so that the number sequence would become 1000, 0001, 1011, 0100, 1101, 1110, 0111 . For many purposes, this is better because it excludes the 0000 case, making all numbers consist of 0's and 1's.

Parity is at its most effective when it is used for small groups of digits, and it can be quite effective in detecting single-bit errors even in eight-bit numbers. It is by no means a method that we could use directly for a digital audio system, however. A simple parity system can only detect the fact that a single bit has changed, it cannot detect which bit, it cannot detect more than one change and it cannot correct any change. As we shall see, however, the principle can be useful when a number can be broken down into groups of bits, and where (as in all digital numbers in binary scale) some bits are much more important than others.

Redundancy is another very important method of detection and correction. In its crudest terms it could mean that each and every number was recorded three times, and the receiving circuits arranged so as to take a 'majority vote' on the correct number if the three versions did not agree. A simpler method would be to record each number twice, with parity included, and exclude the number with a parity error. The simplest and most effective redundancy systems, however, are very wasteful of storage space, and we have to think carefully about what we want to correct. Since the higher order bits in a number convey larger changes in amplitude, it makes sense to employ more redundancy on them than on the lower-order bits.

Like so many aspects of communications by way of digital numbers, redundancy has received close attention from theorists, and is now a very formidable subject in its own right. In this book, therefore, we shall do no more than skim the surface, particularly as regards the complicated systems that have been evolved for compact disc error detection and correction.

Level maintenance is the backstop of support for error correction. If an error has been detected but cannot be corrected, then a level maintenance system discards the number that is in error, and retains the number that was present just previously. The sampling rate for conversion of an audio signal is such that the amplitude difference between two samples will always be very small (even in Aaron Copeland's Fanfare for the Common Man!), so that retaining the previous amplitude value is usually preferable to changing to an incorrect amplitude value, particularly if the change would be a large one. This can even be made the basis of error detection on the basis that no amplitude change from one sample to the next can ever exceed a preset amount. Another possibility here is to use the previous signal with a small dither (noise) signal added, and this gives an even better correction for some types of signals. Yet another possibility is interpolation, in which the amplitude difference between the two previous signals that were not subject to error is added to the signal previous to the error, as outlined in Figure 3.12.

n	n+1	n+2	n+3	...
14605	14622	−22	14655	...
difference = 17		error signalled use 14622 + 17 = 14639		

Figure 3.12 The basis of interpolation. As numbers are read, a circuit finds the difference and stores this. If a new number is signalled as an error, a corrected number is formed by adding the difference (stored from the previous pair) to the last correct value. Even after several interpolations, the waveform shape will not be too seriously affected.

4 Digital to analogue

Simple systems

Conversion from digital to analogue signals must use methods that are suited to the type of digital signal that is being used. If, for example, we were using simple digital amplitude modulation, or even a system in which the number of 1's were proportional to the amplitude of the signal, then the conversion of a digital signal to an analogue signal would amount to little more than smoothing, as Figure 4.1 illustrates. As it happens, it **is** possible to convert a digital signal into a form that can be smoothed simply, and this is

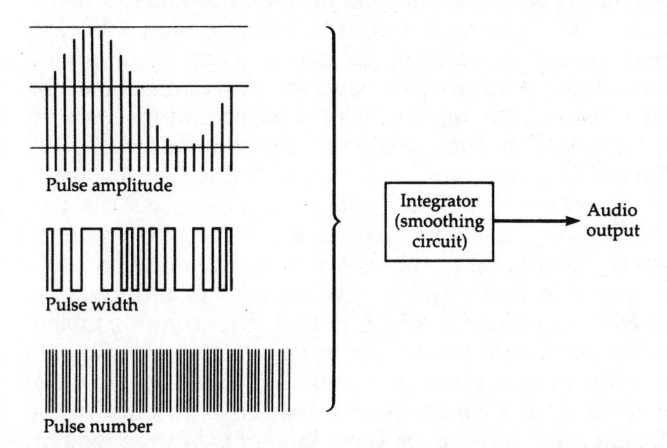

Figure 4.1 Three types of digital modulation that can be converted back to analogue simply by smoothing. None of these is really suitable for digital audio, although the pulse-number system could be very useful if it did not require such a large number of pulses for each number (up to 65535).

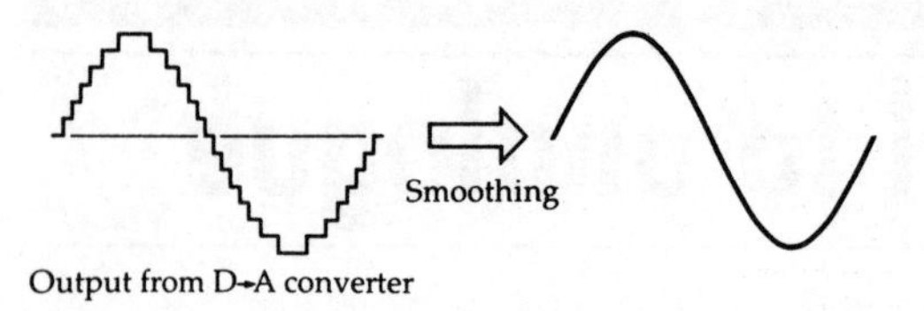

Figure 4.2 The use of smoothing in D-A conversion. A small amount of smoothing will remove the steps from the signal, but a good low-pass filter is better in this respect.

the basis of bitstream methods, as we shall see later. For the moment, however, we shall concentrate on the earlier methods.

Smoothing always plays some part in a digital to analogue conversion, because a converted signal will consist of a set of steps, as Figure 4.2 shows rather than a smooth wave (is any audio signal a smooth wave?). With a sampling rate of the order of 44 kHz, however, the steps are close spaced, and very little smoothing is needed. Even a crude resistor-capacitor smoothing circuit can work wonders with such a wave, and the more complex integrators can reproduce a signal whose amplitude at any point follows closer to the original than is possible when a purely analogue recording method has been used. The faster the pulse-rate of the output of the D-A converter, the easier it is to smooth into an acceptable audio signal.

The pulse code type of digital system does not allow a set of digital signals to be converted into an analogue signal by any simple method, however, which disposes of any dreams of driving a loudspeaker directly from such signals. Remember that the digital signal consists of a set of digitally coded numbers which represent the amplitude of the analogue signal at each sampling point. For conversion, then, we need circuits that will convert each number into a voltage amplitude that is proportional to that number. The alternative, which **would** allow for driving loudspeakers directly, would be to convert each digital number into a set of pulses equal to that number, so that a '1' would give one pulse, and a 65536 would give 65536 pulses. One minor problem with such a system would be that since negative numbers could not be catered for, the path from the smoother to the output stage would need to include a capacitor or transformer to remove the DC level. The alternative would be some kind of bias arrangement. As it happens, bitstream methods bring all this a little closer.

The nature of the conventional conversion requirement is more easily seen if we illustrate it using a four-bit number. In such a

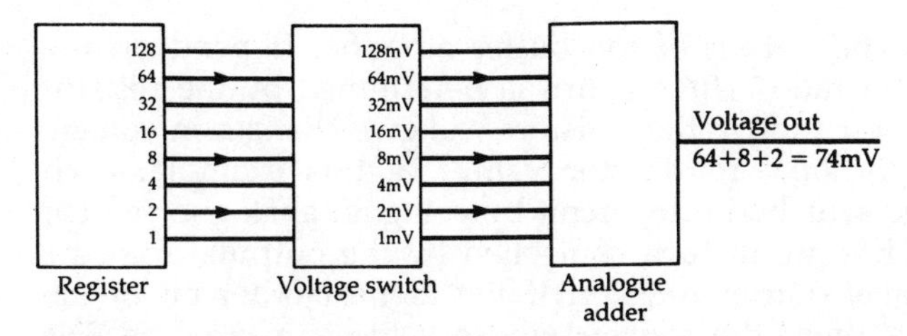

Figure 4.3 The principle of the weighted converter. Each output from the register operates a voltage switch, with the voltages arranged to be in the ratio of powers of two. By adding the voltages, the amplitude of the analogue signal is recovered.

number (like any other binary number), each 1 bit has a different weighting according to its position in the number, so that 0001 represents one, 0010 is two, 0100 is four and 1000 is eight. The progression is always in steps of two, so that if we could connect a register to a voltage generator in the way illustrated in Figure 4.3, each 1 bit would generate a voltage proportional to its importance in the number. In this way, a 1 in the second place of a number would give, say, 2 mV, and a 1 in the 4th place would give 8 mV, with a 1 in the 7th place giving 64 mV. Adding these voltages would then provide a voltage amplitude proportional to the complete digital number, 74 mV in this simple example. This, in essence, is the basis of all D-A converters.

Reverting to a four-bit number, Figure 4.4 shows the basis of a practical method. Four resistors are used in a feedback circuit

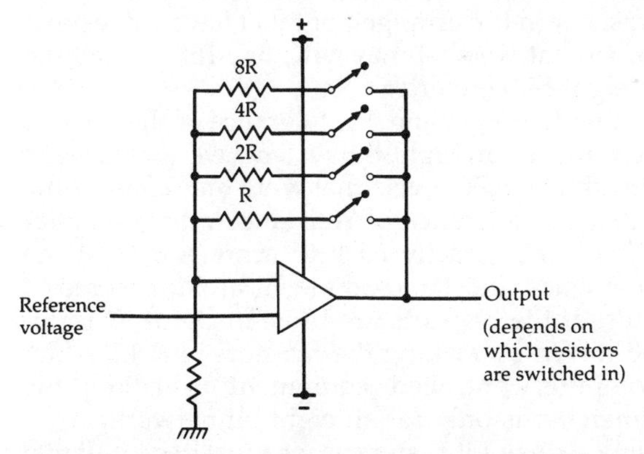

Figure 4.4 The voltage adder stage of a weighted converter, shown for four bits only. The stage is an operational amplifier with switched resistors in the feedback path. The drawbacks of the system are the huge range of resistance values and the requirement for very close tolerance, particularly for the larger values.

which makes the output of the buffer amplifier depend on the voltage division ratio. This in turn is determined by the resistor ratio, so that switching in the resistors will give changes in voltage that are proportional to resistor value. In this example, each resistor can be switched into circuit by using an analogue switch, and the switches are in turn controlled by the outputs from the bits of a parallel output register. If the highest order bit in the register is a 1, then, the resistor whose value is marked as R is switched in, if the next bit is 1, then resistor 2R is switched in and so on.

The result of this is that the voltage from the output of the buffer amplifier (an analogue adder circuit) will be proportional to the size of the digital number. We can at the moment ignore the effect of negative numbers because they can be dealt with by altering the voltage that is switched in with the highest order of bit – the principle of how the converter works is not affected. The attraction of this method is its simplicity – and that is also its main problem.

The resistor switching type of converter is widely used and very effective – but only for a limited number of bits. The problems that arise are the range of resistances that are required, and the need for quite remarkably precise values for these resistors. Suppose, for example, that we considered a circuit in which the minimum resistance that we could use was around 2K. This is a reasonable value if we assume that the amplifier has an output resistance that is not negligible and that we have to use resistance values that will be large compared to the resistance of the analogue switch when it is on. Now each resistor in the arrangement will have values that rise in steps of two, so that we shall have 4K, 8K, 16K . . . all the way to 256K for an eight-bit converter.

Even this is quite a wide range, and the tolerance of the resistor values must also be tight. In an eight-bit system, we would need to be able to distinguish between levels that were one 256th of the full amplitude, so that the tolerance of resistance must be better than one in 256. This is not exactly easy to achieve even if the resistors are precisely made and hand-adjusted, and it becomes a nightmare of difficulty if the resistors are to be made in IC form. The problem can be solved by making the resistors in a thick-film network, using computer-controlled equipment to adjust the values, but this, remember is only for an eight-bit network.

When we consider a sixteen-bit system as we must use for digital audio, the conversion looks quite impossible. Even if we drop the minimum resistance value to around 1K, the maximum will then be of the order of 65M, and the tolerance becomes of the order of

0.006%. The speed of conversion, however, can be very high, and for some purposes a sixteen-bit converter of this type would be used, with the resistors in thick-film form. For any mass-produced application, however, this is not really a feasible method. The requirement for close tolerance can be reduced by carrying out the conversion in four-bit units, because only the highest order bits require the maximum precision.

Incidentally, these converters and the current converters that are described below, all work with positive values only. This means that the digital numbers, though coded in normal form, need to be interpreted differently, and the convention is that the binary zero number (sixteen zeros) will be represented by half of the maximum possible voltage or current, and that positive values correspond to having the most significant bit **set**, the opposite of the digital convention. This amounts to nothing more than the digital equivalent of bias and inversion.

Current addition

The alternative to adding voltages is the addition of currents. We can, of course, regulate currents with resistor networks, and though the values of the resistors that we can use are more acceptable, the requirement for high precision is just as it is for the voltage addition method if we follow the same pattern as the voltage converter. We can, however, consider something rather different. Instead of converting a sixteen-bit number by adding sixteen voltages of weighted values (each worth twice as much as the next lower in the chain), we could consider using 65535 current sources, and making the switching operate from a decoded binary number.

Once again, it looks easier if we consider a small number, three digits this time, as in Figure 4.5. The digital number is 'demultiplexed' in a circuit which has eight outputs. If the digital number is 000, then none of these outputs is high, if the digital number is 001 then 1 output is high, If the digital number is 010, then two outputs are high and if the digital number is 011 then three outputs are high, and so on. If each output from this demultiplexer is used to switch in an equal amount of current to a circuit, then the total current will be proportional to the value of the binary number.

The attraction of this system is that the requirement for precise tolerance is much less, since the bits are not weighted. If one current is on the low side, then another is just as likely to be on

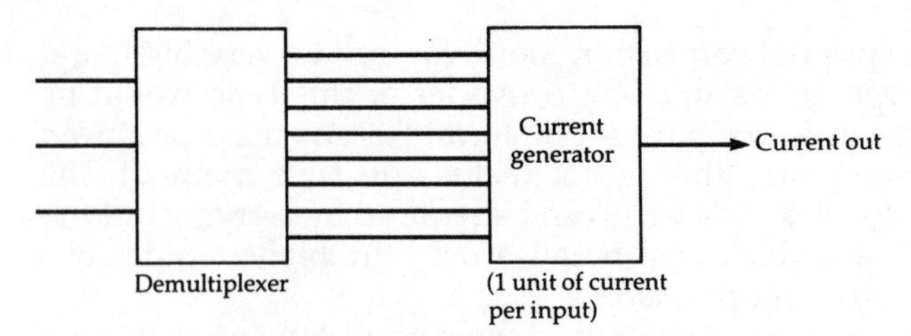

Figure 4.5 Current addition method. The demultiplexer has as its input the binary signals (3 lines in this case) and as its outputs a set of lines. Each binary input causes a corresponding set of outputs to go high, so that a 001 input makes one line high, and a 101 input makes 5 lines high. The demultiplexer outputs are used to operate current generators, each contributing one unit of current to an adder stage (not shown separately).

the high side, and the differences will cancel each other out, something that does not happen if, for example, the one that is low is multiplied by 2 and the one that is high is multiplied by 1024. The effort of constructing 65535 current supplies, each switched on or off by a number in a register, is not quite so formidable as it would seem when you think of what can be done with ICs nowadays. It still amounts to rather more components than could comfortably and reliably be made in IC form when digital audio was in its early stages, though, and compromises are needed to cut down the number of components.

The main compromise is in the number of currents. Instead of switching 65535 identical current sources in and out of circuit, we choose a lower number and make the currents depend on the place value of bits. Suppose, for example, that we settled for 16383 currents, but that each current could be of 1, 2, 4, or 8 units. The steps of current could be achieved by using resistors whose precision need not be too great to contemplate in IC form, and the total number of components has been drastically cut, even if we allow for the more difficult conversion from the digital sixteen bit number at the input to the switching of the currents at the output. This is the type of D-A converter that is most often employed in CD units.

Another possibility is to use a current analogue of the resistor switching method used in the voltage step system, so that the currents are weighted in 2:1 steps, and only sixteen switch stages are needed. This is feasible because we can make current dividers

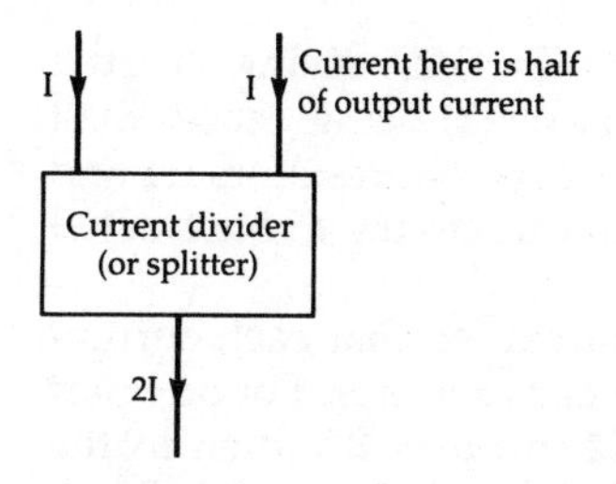

Figure 4.6 The principle of a current divider. If you think of it as a current adder, the action looks simpler.

in IC form, using semiconductors. The principle of a current divider is shown in Figure 4.6 with currents shown as flowing **into** the terminals at the top rather than out – the distinction between a current adder and a current divider is a matter of which direction of currents you are interested in! To act as a precise divider, however, the input currents must be identical to a very close tolerance, and this cannot be achieved by a resistive network alone.

In this form of converter, then, first described by Plassche in 1976, the resistor network is used along with current switching. The principle is shown in Figure 4.7, and is easier to think of if you regard it at first as a current **adder** rather than a divider. The two input currents I1 and I2 form the current I3, and if the resistors R1

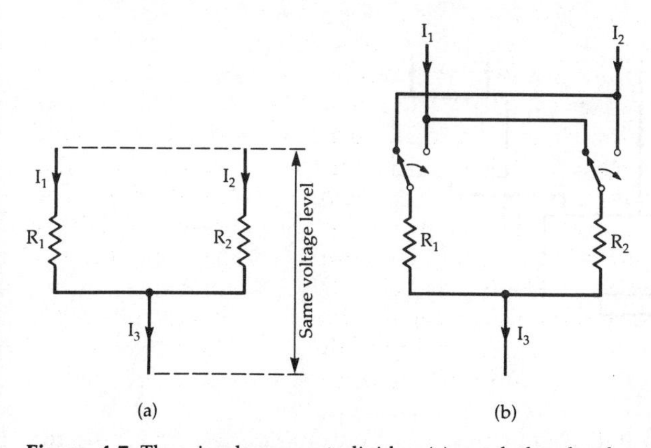

Figure 4.7 The simple current divider (a), and the development of this with switching (b). By switching at a high speed, the division can be almost perfect providing that the clock pulses are at precisely equal intervals, and that the switching transients are smoothed out.

and R2 are precisely equal, then I1=I2=I3/2, giving us the condition that we require of a current (I1 or I2) being exactly half of the other current I3. The snag, of course, is that resistors R1 and R2 will not be identical, particularly when this circuit is constructed in IC form.

The ingenious remedy is to alter the circuit so that each current is switched so that it flows alternately in each resistor. For one part of the cycle, I1 flows through R1 and I2 through R2, then in the other part of the cycle, I1 flows through R2 and I2 through R1. If the switching is fast enough and some smoothing is carried out, the differences between the resistor values are averaged out, so that the condition for the input currents I1 and I2 to be almost exactly half of I3 can be met even if the resistors are only to about 1% tolerance. Theory shows that the error is also proportional to the accuracy of the clock that controls the switching, and this can be made precise to better than 0.01% with no great difficulty.

Now if we consider a set of these stages connected as in Figure 4.8, we can see that the ratio of currents into the stages follows a 2:1 step, and this can be achieved without a huge number of components and without the requirement for great precision in anything other than a clock signal. In addition, the action of the circuit can be very fast, making it suitable for use in the types of A-D converters that were discussed in Chapter 3. The only component that cannot be included into the IC is the smoothing capacitor that is needed to remove the slight current ripple that

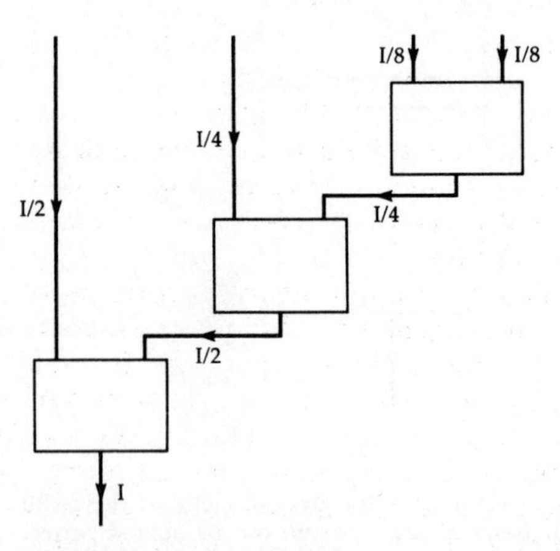

Figure 4.8 The principle of a step-type of current divider, which gives a set of currents that are in the correct ratio for a D-A converter.

will be caused by the switching. This can be an external component connected to two pins of the IC.

The level of conversion is another point that needs to be considered. If we take a voltage conversion as an example, then the amplitude of the signal steps that we use will determine whether the peak output is 10 mV, 100 mV, 1V or whatever. The higher the peak output, the more difficult is the conversion because switching voltages rapidly is, in circuit terms, much simpler when the voltage steps are small, since the stray capacitances have to be charged to lower levels. On the other hand, the lower the level of the conversion, the greater the effect of stray noise and the more amplification will be needed. A conversion level of about one volt is ideal, and this is the level that is aimed at in most converters.

Conversion problems

The problems that arise with the conventional type of D-A converter are closely connected with the nature of the digital signal and the ever-present problems of precise current generation. A 16-bit D-A converter requires the use of 16 current sources, and the ratio of a current to the next lower step of current must be exactly 2. For a conventional type of current generator system in which the current source provides current I, the most significant bit will be switching a current equal to I/2 and the least significant bit will be switching a current equal to I/65536. With all switches open (all bits zero) the current will be zero; with all switches closed (all bits 1) the current will be equal to I – I/65536. Note that the current can never be equal to I unless the number of bits is infinite; in this example I/65536 is the minimum step of current. The value of I/2 is taken as being the current corresponding to zero signal, so that zero is being represented by 1000000000000000 – current converters cannot deal with negative currents.

Maintaining the correct ratios between all of these currents is virtually impossible, though many ingenious techniques have been devised. In particular, this scheme of D-A conversion is very susceptible to a form of crossover distortion, when the zero current level changes to the minimum negative current. In digit form, this is the change from the current I/2 (represented by the digital number 1000000000000000) to the lowest negative value which is the sum of all the current from I/4 down to I/65536, corresponding to the digital number 0111111111111111. This step of current ought to be small, equal to I/65536, but if any of the more significant current values are incorrect even to the extent of only 0.05%, the effect on this step amount will be devastating – a rise of seven times the minimum current instead of a fall of the minimum step amount, for example.

This amount of error is virtually unavoidable, and it will cause a cross-over error to occur each time the signal passes through the zero level, corresponding to a current of I/2. In addition to this cross-over problem, all converters suffer from 'glitches', which are transient spikes that occur as the bits change. These are caused by small variations in the time when switches open and close, and they also are most serious when the signal level is at its minimum, because they cannot be masked by the signal level, and the number of bits that change is at its greatest when there is a change from zero (I/2 current) to a small negative value. Both of these problems are answered by the adoption of bitstream methods.

Before conversion

The conversion of the digital signal into analogue form is just one part of a digital audio receiving system. Before the signal is converted it will have to be error checked and corrected, making use of whatever additional bits have been sent with each number and with each group of numbers. We looked in Chapter 3 at the outline of methods that can be used to detect and correct errors, and in more detail at the simplest method of error correction, the parity bit. The use of a parity bit, though an acceptable method for some systems like serial data communication through wires, is not nearly good enough for digital audio systems, and we have to consider the use of more advanced methods which have been developed from computing. Many of these methods have been known for some time and have featured even in comparatively low-priced 'home' computers of the past as error correction systems for the recording and reproduction of digital signals using ordinary cassette tape. The most important of these systems is called Cyclic Redundancy Checking, or CRC, and a simpler process which is easier to understand is the checksum method, which we shall look at first.

The checksum method, as the name suggests, is based on adding the digital numbers over a set. Most digital audio systems work, as we shall see, with a 'frame' that consists of a set of numbers along with synchronizing and error-correcting signals.

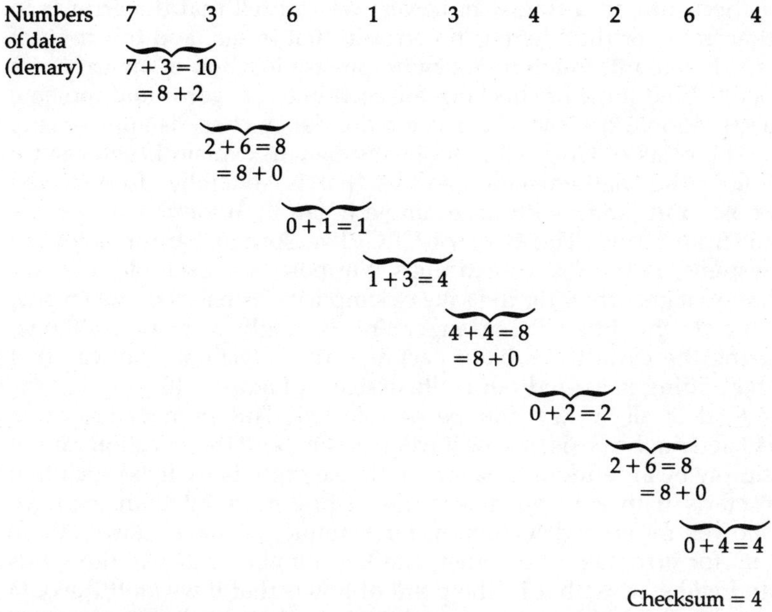

Numbers of data (denary): 7 3 6 1 3 4 2 6 4

$7 + 3 = 10 = 8 + 2$

$2 + 6 = 8 = 8 + 0$

$0 + 1 = 1$

$1 + 3 = 4$

$4 + 4 = 8 = 8 + 0$

$0 + 2 = 2$

$2 + 6 = 8 = 8 + 0$

$0 + 4 = 4$

Checksum = 4

Alternatively: Sum = 36 = 8 × 4 + 4 remainder

Figure 4.9 The checksum method illustrated with 4-bit numbers. The numbers are added, ignoring any carry output of the most significant bit, and the remainder used as the check number which is also transmitted. Repeating this at the receiver should give a matching checksum. In this example, using 4-bit numbers means that the carry is for the number 8, so only the difference between the sum and 8 is carried forward on each addition.

Suppose, for example, that we have 12 numbers in each frame. To make a checksum, we add up the twelve numbers, ignoring any overspill so that the sum will still fit into 16 bits. Figure 4.9 shows an example carried out using four-bit numbers. The checksum number is then recorded as part of the frame. At the receiving end, the numbers are summed in exactly the same way, and compared with the checksum. If there is any disagreement then there has been an error somewhere along the line.

This type of error detection is very much more sensitive than parity, because it is much less likely that a change of 0 to 1 in a number will be balanced by a change from 1 to 0 in exactly the same bit position in another number. The main problem of using a checksum is that it does not indicate where the error is. By using a checksum on a frame, however, we can tell that the error is in that frame or that there is no error in that frame, and this in itself is valuable information. For more precise location of an error, we need some form of checking for each byte or word, and in some applications, the use of a cyclic redundancy check is appropriate.

The basis of CRC is by no means easy to explain if you cannot follow the mathematical analysis that is generally used in text books, but along with an example it can be understood in more concrete terms. The basis of CRC, like so many error detection systems, is the use of extra bits. Suppose, for example, that we have a signal that, for the sake of simplicity, consists of only 5 bits. To code this in a CRC arrangement we might want to use 8 bits, using the extra three bits to act as error detection. The way that this coding is carried out is illustrated in Figure 4.10.

First of all, a 'key' number is selected. This number has to be chosen on the basis of how it will perform, and the selection cannot simply be at random. The size of the key number will be such that it starts with a 1 and makes use of one more bit than we have allowed for error detection. If, for example, we have allowed three bits for error detection, then the key number will use three bits and will start with a 1. The point of this is that if we shift the data number left by three places and then divide it by the key number, the remainder after division will be a number that can be 0 (no remainder) or anything up to three bits of 1's. For example, if we have allocated 3 bits, and we use 7 as the key number, then the maximum possible remainder is 6, 110 in binary.

Having carried out the division, the remainder is then placed in the lower vacant places of the complete number, which now

Data: 01101101 (denary 109)
Key: 101 (denary 5)
Remainder of division: 100 (109/5 = 204/5 4 remainder)
Transmitted as: 01101101100
On reception, regroup as 01101101 and 100
01101101 divided by 101 gives 100. Data has no errors.
If data divided by key is not equal to remainder, errors exist.

Figure 4.10 The principle of cyclic redundancy checking requires more bits to be used. In this example, the 8-bit data number is divided by a 'key' number of three bits, and the remainder obtained. The data is shifted left by three places and the remainder added. At the receiver, another check of the same type should give the same remainder.

consists of data bits and remainder bits. Now if this is transmitted without errors, the data part of the number can be tested by dividing it again by the same key. If no errors have occurred, then the same remainder should be obtained. If, however, there has been an error in the transmission of that number then the same remainder will not be obtained. If the key is a suitable one, then altering the data number so that the correct error number is obtained may restore the correct data. This is not foolproof, but works well enough if combined with rejection of any large changes to be a very useful correction system.

The use of cyclic redundancy checking as such can be elaborated, but for digital audio consumer systems, the use of checking on each word takes up too much space and time. A better method is to use a different form of coding, not the simple binary number, and that's a system that we shall look at in Chapter 6 when we consider the established CD system. First, however, we need to look at two refinements to the system which have arisen since the first CD players started to emerge.

Oversampling

We saw in Chapter 2 that sampling creates a set of pulses which are still amplitude modulated and which correspond to the audio signal plus a set of sidebands around the sampling frequency and its harmonics. After the D-A converter has done its work, this is the signal that will exist at the output and as we have seen, it requires low-pass filtering to pass only the audio portion up to 23 kHz and reject the higher frequencies.

This, however, calls for the use of filters with a very stringent specification, and such filters have an effect on the wanted part of the frequency range which is by no means pleasant. A simple way out of the problem would be to double the sampling rate using, for example, 88.2 kHz for a CD system rather than 44.1 kHz. This is not possible at the recording end of the chain, and since A-D converters are stretched as it is to cope with a rate of 44.1 kHz, it is not possible to double the number of samplings at the A-D converter.

What **can** be done, though, is to add pulses between the output pulses from the A-D converter, Figure 4.11. This creates a pulse stream at 88.2 kHz, and so makes the frequency spectrum quite different, Figure 4.12. The lowest sideband is now at about 68.2 kHz, well above the 20 kHz limit of the audio signal and easily

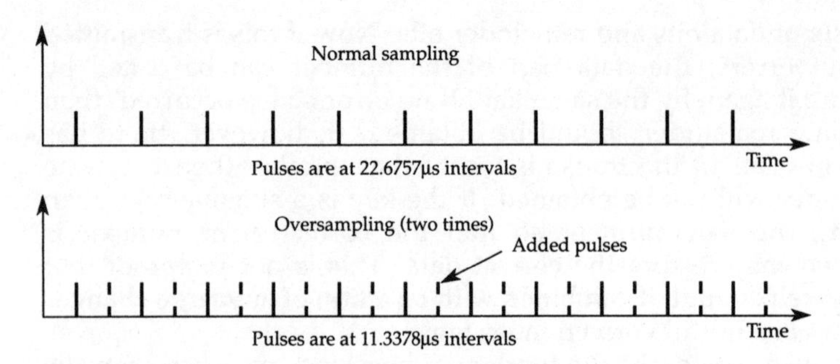

Figure 4.11 Adding pulses that lie between the existing sampled pulses will halve the time between pulses in the signal, providing a two-times oversampling.

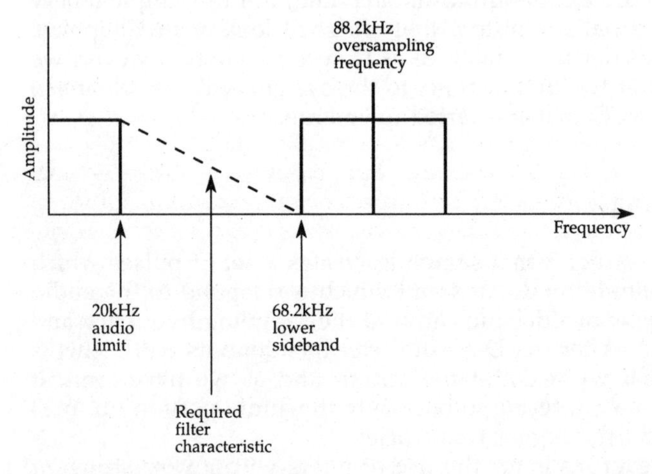

Figure 4.12 The spectrum that arises from two-times oversampling - the gap between the audio signal and the lower sideband of the oversampling frequency is much greater, allowing a simpler and more effective filter to be used.

filtered out. In addition, if the added pulses are midway in amplitude between the original pulses, **interpolated** values, the conversion from pulses to smoothed audio can be considerably smoother, just as if the real sampling rate had been doubled. Oversampling of four times or even more is now quite common on conventional CD systems, but the whole technology of conventional oversampling D-A conversion has been overturned by the adoption of single-bit (more correctly, single-level) conversion methods, of which bitstream is a well-known example. The combination of oversampling and bitstream is now the conventional system for high-quality CD players.

Bitstream methods

The most significant development in digital audio in recent years has been the use of bitstream technology. The output of a bitstream converter is not the voltage or current level that is the output of a conventional D-A converter but, as the name suggests, a stream of 0's and 1's in which the ratio of 1's to 0's represents the ratio of positive to negative in the analogue signal. For example, a bitstream of all 1's would represent maximum positive voltage, one of all 0's would represent maximum negative voltage, and one of equal numbers of 1's and 0's alternated would represent zero voltage.

The enormous advantages of bitstream conversion are that only one current generator is required, there are no ladder networks needed to accomplish impossibly-precise current division, and the output signal requires no more than a low-pass filter – the bitstream signal is at such a high frequency that a suitable filter is very simple to construct. The disadvantage is that the system must at this point handle very high frequencies – of the order of 11 to 33 MHz for the bitstream converters now being used in CD players. The essence of the bitstream converter is that instead of using a large number of levels at a relatively slow repetition rate, it uses very few levels (2) at a much higher repetition rate. The rate of information processing is the same, but it is accomplished in a very different way.

The principles are surprisingly old, based on a scheme called delta modulation which was developed for long-distance telephone links in the 1940's and 1950's. The normal pulse-code modulation system of sampling and converting the amplitude of the sampled signal into a binary-coded number has well-known drawbacks because of the noise that is generated because of the quantization process. This noise can be forced into higher frequency bands by using higher sampling rates, but since the quantization is fixed at 16-bit at the manufacturing side of CD, nothing can be done to alter the quantization noise that is recorded onto the CD.

Other digital systems, however, can use a scheme called delta modulation, in which the *difference* between samples is converted into binary numbers. If the rate of sampling is very high, the difference is very small, and it can be reduced to one bit only. A signal of this type can be converted back to analogue by using an integrator and a low-pass filter. Because of the very severe distortion that occurs when a delta modulator is overloaded, the sigma delta system was developed. This uses sampling of the signal amplitude rather than differences, and forms a stream of pulses by a feedback system, feeding the quantized signal back to be mixed with the incoming samples in an integrator. This results in a noise signal which is not white noise, spread even over the whole range of frequencies, but coloured noise, concentrated more at the higher frequencies. Because of this characteristic, this type of modulator is often termed a 'noise-shaper. This is the type of technology, used on other digital systems, which has led to bitstream and other forms of D-A converters in which the number of bits in the signal is reduced.

Current bitstream converters consist basically of an oversampling stage followed by the circuit called the *noise-shaper*. The oversampling stage generates pulses at the correct repetition rate, and the noise-shaper uses these pulses to form a stream of 1's and 0's at the frequency determined by the oversampling rate. The important part of all this is the action of the noise-shaper.

Figure 4.13 shows the principles of a noise-shaper, which consists of two adders, a converter, and a delay – the amount of the delay is the time between pulses. If we concentrate on the converter for the moment, this is a circuit which has a very fast

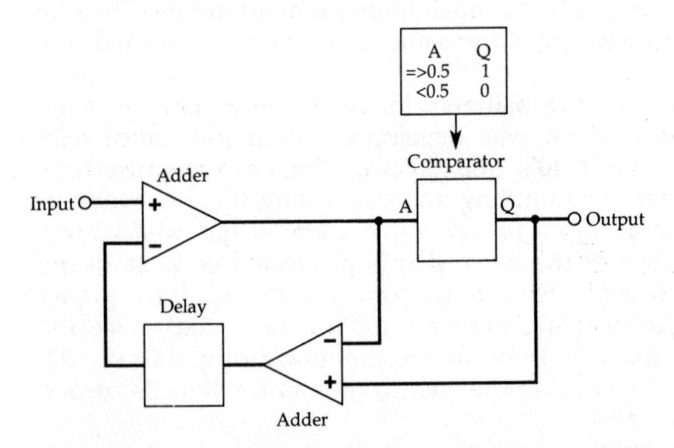

Figure 4.13 A block diagram for a form of noise shaper, using two adders, a comparator and a delay. The adders are digital adders, and the comparator has a special truth table shown in the diagram.

response time and which will give an output of 0 for numbers corresponding to less than 0.5 of full output, and an output of 1 for numbers corresponding to more than 0.5 of maximum level. The generation of the bitstream is done rather like the conversion of a denary number into binary, by successive comparisons giving a 0 or 1 and feeding the remainder of a comparison back to be subtracted from the next (identical) input signal. Note that 0 and 1 from the converter corresponds to 0 or I_{max} current values in a conventional D-A circuit rather than the digital 0 and 1 of a number.

Imagine a number at the input causing either a 0 or a 1 signal to be developed. The second adder works with one inverted signal, so that its output is the difference between the original signal amplitude number and the maximum signal (output 1) or minimum signal (output 0). This difference is delayed so that it appears at the input adder while the same oversampled copy of the input is still held there – the amount of delay will fix how many times this action can be carried out on each input. If the error signal is negative it will be subtracted from the input, otherwise it is added, and the result is fed to the converter. This again will result in a max or min output signal and an error difference, and the error difference is once again fed to the input adder to be added to the new input and so to be converted. At each step, the output is a 1 or a 0 at many times the pulse repetition rate of the oversampled signal, and with the ratio of 0's and 1's faithfully following the amplitude corresponding to the input numbers.

Figure 4.14 illustrates the principle. The input number represents an amplitude of 0.7 and this input will cause the converter to issue an output of 1. The difference is −0.3, and when this is added to the next identical input it gives a net input of 0.4 which causes a 0 output with an error of +0.4. The diagram shows the successive steps which end when the net input to the converter is

IN	OUT	FEEDBACK
0.7	1	−0.3
0.7 − 0.3 = 0.4	0	+0.4
0.7 + 0.4 = 1.1	1	+0.1
0.7 + 0.1 = 0.8	1	−0.2
0.7 − 0.2 = 0.5	1	−0.5
0.7 − 0.5 = 0.2	0	+0.2
0.7 + 0.2 = 0.9	1	−0.1
0.7 − 0.1 = 0.6	1	−0.4
0.7 − 0.4 = 0.3	0	+0.3
0.7 + 0.3 = 1.0	1	0.0 END

Bitstream contains seven 1s and three 0s; 70% full amplitude or 0.7V

Figure 4.14 Illustrating how a voltage level in digital terms is reduced to a set of 0 and 1 bits to form a bitstream by the action of the noise shaper.

1 so that there is no error signal from the output – this ends the conversion which in this example has taken ten steps – implying that the output frequency will be ten times the input frequency. The output is a stream of 7 1's and three 0's, corresponding to a wave that is 70% of full amplitude, an amplitude of 0.7 of maximum.

As might be expected, practice is rather more complicated than this simple example suggests, and some of these complications are noted in Chapter 6, dealing with the application of bitstream to recent CD players.

5 Studio digital methods

Digital tape

Studio recording techniques now concentrate on digital tape methods rather than on analogue, so reducing the problems of quality in tape mastering. The main problem, which is a very old one that has always dogged the fortunes of tape recording, is that the magnetizing of a tape is not a linear process. The graph of retained magnetism plotted against amplitude of magnetizing current is 'S' shaped, and at very low values of magnetizing current, the tape is not magnetized at all. The traditional way of partially overcoming this has been to add ultrasonic bias to the recording head so that the audio signal was superimposed on the peaks of a high frequency signal. By arranging for a suitable amplitude of bias, these peaks could be placed in a region of the tape characteristic that was reasonably linear. The linearity obtained in this way was just acceptable – but only just.

The main advantage of digital recording for a tape medium is obvious. Since a signal consists only of 0 or 1 digits, there is no requirement for linearity, so removing the greatest problem of tape recording. Compared to this, all other considerations become minor ones because all analogue tape recording is a struggle to keep reasonable linearity while trying to record a good dynamic range of sound amplitude without overloading the tape. In the past, this has involved very elaborate companding circuits whose contribution to the fidelity of sound has been acceptable in some cases, dubious in others. Since digital reproduction requires no variation of amplitude of signals, the problems of tape overload and saturation also disappear.

Now that digital recording techniques are well established, it's difficult to remember just what immense problems tape always represented as an analogue medium. While distortion figures of 0.1% were commonplace for amplifiers, magnetic recording was striving to get below 1% distortion, and even when this had been achieved, to stay there consistently was quite another matter. The situation was such that several of the smaller recording companies who could not make a colossal investment in modern techniques were reverting to direct disc cutting to avoid the problems of tape mastering and the temptation that all sound engineers have to 'doctor' the sound while it is on the tape.

Before we get too euphoric, though, it's as well to remember the difficulties that faced the pioneers of digital tape recording. The first (and main) problem is that a digital signal requires a much higher recording density than an analogue one. If we stay with the CD standards of around 44 kHz sampling rate, and recording 16 bits at each sample, plus allowance for extra error-checking and correcting bits, then the number of 1 or 0 signals per second becomes very formidable, of the order of several megahertz. This takes us into video frequency rates, and digital audio would probably not have been developed at anything like the rate that was possible if video recording had not smoothed a path. Digital recording is, if anything, rather easier than video recording because the video signal is not digital. In order to record video signals we have to convert them to frequency modulated signals of constant amplitude, making them rather like digital signals.

Even allowing for the fact that the digital signal can be easily recorded, using, for example, one direction of magnetization for 0 and the opposite direction for 1, the rate of recording is still very high. The bandwidth of a digital recorder needs to be about 30 times as much as is needed for analogue recording – and in the days of analogue tape mastering it was almost as much a struggle to achieve adequate bandwidth as it was to achieve acceptable linearity. This, however, is the only really serious problem, and all other considerations point to digital recording as a preferable method. There are, in addition, some useful bonuses, like the ability to interleave two or more stereo channels into one track, or to interleave audio (digital) signals with control signals.

There are quite a large number of methods of recording digital signals on tape (or other magnetic media) of which the NRZ (non-return to zero) system is the simplest. NRZ recording uses two directions of magnetization, referred to for convenience as positive and negative though there is no connection with voltage positive and negative, to represent digital signals with positive meaning 1 and negative meaning 0. The main problem of NRZ methods is that if the bits of a signal are unchanged for a long period the signal that is recorded will be the equivalent of a DC signal, and will suffer distortion, since magnetic heads used on replay will respond to the rate of **change** of magnetization on the tape, not the extent of magnetization.

NRZ is less of a problem when data has been coded as FM signals, so that it is a method that is used for video recorders when they are working with digital signals that have been frequency modulated. For other purposes, NRZ is not favoured, and it was replaced many years ago for digital signal purposes by more modern methods. There is a modification of NRZ which adds an extra clock bit so that each data signal consists of a clock signal followed by a data signal, with the clock signal always in the opposite direction. This ensures that each data bit is represented by a change in magnetization, and so avoids the problems of long strings of identical bits. This type of code has been used for recording sound on the Video-8 system when the recorder is switched to sound-only; the bits are then frequency-modulated.

A much-used method for computing is known as MFM, modified frequency modulation. This is used extensively on computer magnetic discs, though it is steadily being replaced by other methods. Referring to the diagram, Figure 5.1 each bit is taken as occupying a time called the *bit cell*, and changes between positive and negative directions of magnetization occur either in the middle of the cell or at the start. Where there is a zero in a cell, there is no change in the direction of magnetization in the middle of the cell. Where there is a 1 in a cell, the direction of magnetization occurs in the middle of the cell time. There is always

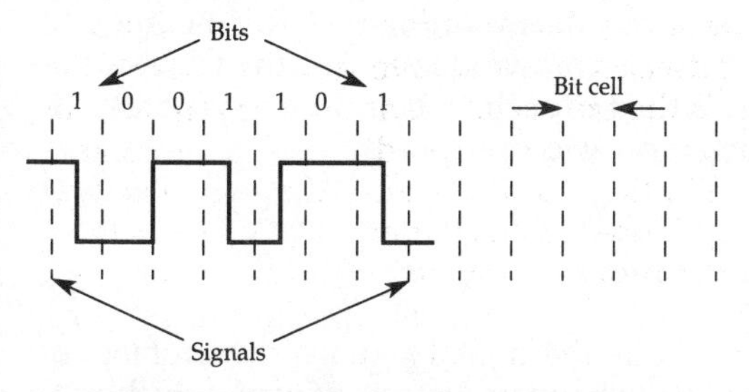

Figure 5.1 Illustrating MFM recording in which each 1 bit causes a transition (0 to 1 or 1 to 0) in the middle of the bit cell and two adjacent 0's will cause a transition at the start of the bit cell.

a change of direction of magnetization at the start of any cell in which a second or subsequent zero occurs.

Since the MFM coding is by change of direction rather than by absolute direction, there is no problem with long strings of 1's or 0's. On the other hand, the coding and decoding is more elaborate and it is necessary to keep a precise clock rate in order to be able to distinguish a change of magnetization that occurs at the start of a cell as compared to one that occurs in the middle. In addition, the use of MFM requires changes to occur at frequent intervals, as close as half a bit cell, and this limits the density with which the signals can be packed on to the magnetic tape or disc surface.

Other methods depend on code conversions, and of these the system of most interest is the EFM (Eight to Fourteen) system as used for compact disc – since the principles of so many other coding systems follow much the same pattern only EFM will be explained here. The same system, used for computer hard discs is known as the RLL (run-length limited) system, and it allows tighter packing of data on to the disc area as compared with the older MFM.

Using EFM, the data bits are dealt with not individually but in blocks of eight, corresponding to the byte unit used so much in computing. Each set of eight bits is then converted to a 14-bit pattern, using a conversion table (a ROM which will give the appropriate 14-bit output for an 8-bit input). A collection of 14 bits can represent 2^{14} states of 1 and 0, a total of 16384 possible arrangements. Since 8 bits can be used to represent 256 possible states, the conversion to 14 can be done in such a way that each 14-bit code has no string of identical digits of more than 7 nor less than 2. This simultaneously deal with both requirements – no long sequences and no 1010 types of transitions. In computing, this is often referred to as the RLL(2,7) system. Once modulated, these 14-bit units can be recorded by ordinary NRZ methods to give tight packing and better error-avoidance. For the CD form of EFM, three extra bits are added to each 14-bit unit for synchronization and low-frequency suppression purposes.

The tape bandwidth problem

No matter what we do to code the digital signals, the problem of bandwidth can be solved in only one satisfactory way, which is to increase the rate at which the tape passes the recording head. This can be done by using a stationary head and moving the tape rapidly past the head, as used in early types of video and digital audio recorders. More useful systems use multitrack heads, the basis of the S-DAT system, or the more familiar video recorder technique of using rotating heads which cross the slow-moving tape at an angle. For studio recording, the use of fixed head machines is not such a disadvantage as it would be for domestic use (the requirement for very long tapes in large reels, for example), and fixed head machines were the first to be developed and used. One mitigating point about a fixed-head machine is that it readily allows more than one track to be recorded, since studios normally want at least 24 tracks. It also allows for easier cut-and-splice editing, though editing systems for other forms of recording are by now well established.

The critical quantities, as far as any tape recording system is concerned, are the minimum recorded wavelength and the maximum frequency. The minimum recorded wavelength is the minimum distance along the tape on which one complete wave can be recorded. This quantity depends on the gap in the recording head and will be in the range 1 to 5 μm (1 μm = one millionth of a metre, one thousandth of a millimetre, about 40 millionths of an inch). Suppose that the tape is moving over the head at 5 cm per second (taking a convenient figure rather than an actual one) and the minimum wavelength is 5 μm. With this figure of minimum wavelength, we can record 1000000/5 = 200,000 cycles in one metre of tape, and so get 200,000/20 = 10,000 cycles into 5 cm of tape. At a speed of 5 cm /s therefore we can record 10,000 cycles in a second, giving a bandwidth of 10 kHz.

For recording frequencies in the region of 4 MHz, then, with the minimum recorded wavelength of 5 μm, we are going to need tape speeds of around 20 metres per second. Even for studio use this is unacceptable, and studio digital tape recorders in the past have used simpler digital methods that allowed the bandwidth to be reduced to less than 1MHz. By further 'trickery', such as splitting data between different channels on the tape, tape speeds as low as 38 cm/s could be used in fixed-head recorders.

The use of rotating head recorders, though it brings problems of switching and synchronization, greatly relieves the problem of tape speed, and allows tape speeds comparable to the speeds of domestic tape cassettes to be used. In addition, it allows the PCM systems, comparable to the CD system, to be used rather than other digital modulation systems. The acknowledged pioneer of rotating head recorders with PCM are Nippon Columbia of Tokyo, who in 1972 were able to demonstrate a PCM audio system operating on a professional video tape machine (domestic video recorders did not exist at that time). Since that time, many other forms of digital audio studio recorders have been developed – and so also has the CD system. As you might expect, this has raised problems of compatibility.

One of the reasons for the compatibility problem has been that both video recording and digital audio recording have been developing very fast and in parallel with each other. Quite apart from digital methods, there has always been a lack of uniform standards in regard to recording throughout the world. As video recording developed, incompatible standards developed there also – in the domestic VCR field we have seen the example of VHS, Betamax and the various Philips standards competing. It is hardly surprising, then, to find that there are at least five major incompatible digital recording systems in use at the present time for studio recording on reel-to-reel tape.

The enormous success of the collaboration of Philips with Sony in developing a single world-wide standard for compact disc has had some effect, however, and manufacturers are moving towards studio systems that have at least some compatibility with CD. Fixed head machines are not so common now, though they still have an advantage in allowing easy splicing of the razor-blade and block variety. This can be an important consideration for a small studio. In addition, a fixed head machine can employ another head for monitoring the recorded signal, something that is not generally possible with the rotating head type.

The Sony PCM 1620 and 1630 machines were among the first rotary head machines used for studio recording, and are still widely used though rotary head machines are now available from a number of other suppliers. Since the modern R-DAT (rotary head digital audio tape) cassette systems also use rotating heads and work with digital signals that are compatible with CD standards, we can expect to find that studio recorders will fall into line. The R-DAT cassettes have been available for some time, but their sales have been hindered by objections from the recording studios who feel that a recording system that could make perfect tape copies from CDs would undermine the market for CDs. The answer, as the computer software business has shown, may be to sell CDs at a price that is competitive with the price of a blank R-DAT tape. The problem is probably not so great as it might seem. The urge to transfer LPs to tape was for a large part motivated by the rapid deterioration of the LP as it was played several times over. Since a CD has virtually an unlimited life, there is little point in taping it except for use in a car fitted with conventional cassette equipment. One wonders if the objections would have been so strong if R-DAT had been developed in Europe and the major CD suppliers had been in Japan. As it happens, events have thrown a cloud of doubt over the whole future of R-DAT; we shall return to this in Chapter 7.

Rotary head techniques

We have made a lot of mention of rotating head recorders in this chapter, but to the reader who has taken no interest in video cassette techniques, rotating head recorders are as unusual as digital recording itself. The following section is an introduction to the ideas of rotary head tape recording for the benefit of the reader whose experience is almost entirely based on conventional analogue audio equipment. If you have experience with video recording then most of what follows will be familiar, though the use of rotary head recording with digital audio signals is not identical in pattern to that used for videocassettes. If you need a more comprehensive treatment of rotating-head video recording then you will find a book by Eugene Trundle, 'Television and Video Engineers Pocket Book' very helpful; (see Appendix 1).

To start with, the reason for adopting rotary head recording for videocassettes was that a manageable cassette can contain only a limited length of tape, and with fixed head recording such a quantity of tape could provide only a few minutes of playing. To put it into perspective, the record/play speed for a fixed head recorder would need to be of the same order of speed as the fast-wind of a modern videocassette recorder. The solution that has been adopted for videocassette recording, and also on some other videotape machines, is to use revolving tapeheads whose speed relative to the tape can be very high, allowing the tape itself to be moved at a comparatively slow pace, about 1.873 cm/s (0.737 inches/s) on the Betamax type of recorder.

The usual technique is to have two tapeheads revolving fast enough to give a speed between head and tape of between 5 and 7 m/s depending on the type of system – the higher speed is used by the Betamax system. The path of the tape is not in the plane of rotation of the heads, but at a small angle of about 5° so that each head will cross the tape so that it lays down a diagonal track shown in Figure 5.2 with the angle considerably exaggerated to make the tracks more obvious. The angle around which the tape is wrapped

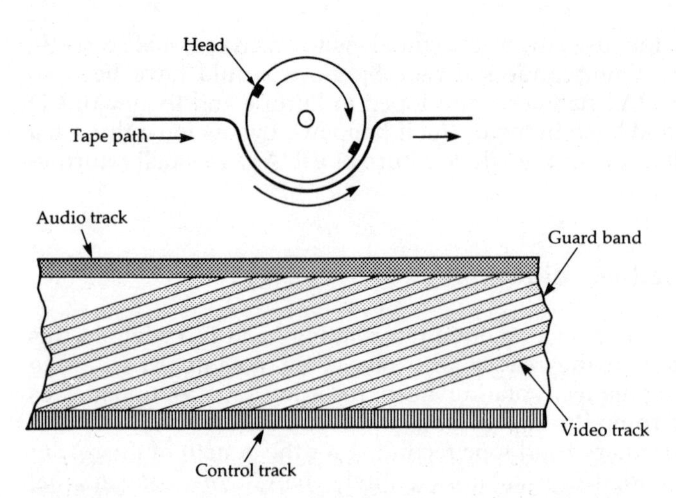

Figure 5.2 The principle of rotating-head video recording. The tape is wrapped around a guide so that a drum containing the two heads can pass across the tape at a small angle. This lays down a pattern of diagonal tracks separated by an unrecorded 'guard band'. The overlap between the heads is not used, because only synchronizing signals are transmitted in this time, so that the edges of the tape can be used for audio and for control signals.

is about 186 degrees. The movement of the tape itself is such as to arrange for a small gap, the guard band, to exist between successive video tracks. Each guard band therefore separates tracks that have been recorded by alternate heads. The edges of the tape pass over stationary heads, one of which records the audio signals on one side of the tape, with the other recording synchronizing and control signals on the other edge. In the older design of videocassette recorder, the sound quality is very poor because the audio signals are being recorded with a stationary head and a tape speed that is very low even by audio cassette standards. More modern units use an additional rotating audio head, following the same methods that have evolved into the R-DAT system. The video signals are separated into sections and frequency-modulated on to a carrier with the colour (chroma) signals at a lower frequency band than the luminance (brightness) signals. The use of frequency modulation minimizes the need for linear tape response and also allows for speed jitter by recording signals which when retrieved can be used to create a local clock signal. In addition, the two recording/replay heads are placed at different azimuth angles to reduce interference between adjacent tracks, and by storing and comparing signals, the effects of crosstalk can be minimized.

At first sight, you might think that this well-tried system could be adopted in a perfectly straightforward way to digital audio recording, because the digital audio signals are not quite so demanding. The problem is that audio signals are continuous, whereas video signals are not. A video signal consists of a set of waveforms which repeat at 20 ms intervals, the field time. Each field of a video signal corresponds to a set of lines being drawn on the screen of the receiver, and two fields make up a complete picture. The fields are interlaced, meaning that the odd-numbered lines of a picture are in one field and the even numbered lines in the next field. The important point here is that there is a time gap between fields, large enough for the receiver cathode ray tube to move the scanning spot back to the top left hand of the picture. This time is the field synchronization period, and takes the time of twenty lines in each field.

This field synchronization period contains only synchronizing pulses, not picture, and these pulses contain timing information, not picture signal. If the rotation of the heads is suitably governed, then, and the amount of tape wrap is correct, the crossover from one head to another can be arranged to take place during the field synchronizing interval rather than at a time when picture information is being sent. One tape track from one head, in other words, records or contains the picture signal data of one complete field. The field synchronization pulses occur at regular intervals and can be replaced by clock pulses from an oscillator.

This discontinuity of video signals makes the unavoidable problem of the changeover from one head to another rather insignificant when a TV signal is being recorded. There is, however, no such natural break in an audio signal, analogue or digital, and if we are going to use a two-head system then we have to create a break. This is the basis of all systems that use standard videocassette recorders for digital audio, and also the R-DAT digital audio cassette system.

Creating a break is done by time-compression of the digital pulses that make up a 'frame' of signal. This signal 'frame' is entirely artificial, simply the number of digital signals that will fit into the time that a head requires to scan its path across a piece of tape. By time compression, I mean that the number of digital pulses that for one frame are sent to the recording head (or received from it) in a time that is shorter than their natural period. This allows for a gap in the interval when the scanning of the tape

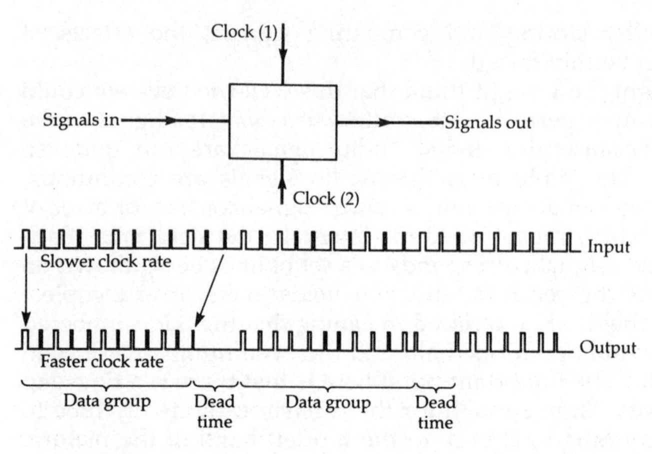

Figure 5.3 Data compression used to make artificial gaps in data for use with rotating head audio recorders. The signals are placed into memory with one clock speed, and read with a higher clock speed, with a gate used to interrupt the reading clock at intervals. The illustration shows pulses grouped in twelves, but larger numbers are used in practice. At the receiver, the reverse process restores the original clock rate and the original pulse rate.

is handed over from one head to the other, after which another frame of signals can be sent or received.

This time compression is achieved by the use of computer-style memory. The signals in digital form taken from the processing circuits are loaded into memory continuously, but the memory is read only at intervals that are separated by a gap, Figure 5.3. This needs more than a simple serial shift register to accomplish, because the bits of memory are read faster than they are written (on the recording side), so that the memory needs to accommodate a complete frame of signals plus the signals that arrive in the time between frames. The solution is the use of random-access memory (RAM) as employed in computers – and if this is not a familiar system we shall have to explain it further here.

Unlike memory based on shift registers, RAM uses a set of flip-flops in which any one of the set can be selected without the need to select any other in a sequence – we can, in other words select at random, hence the name. In such a memory, selection is done by using a binary number, called an address, to locate each unit of memory, so that each unit has its own unique address number. On the type of RAM used for computers, reading and writing never takes place at the same time, so the memory can be switched to either function so as to allow one set of lines to carry data either to the memory or from it. For digital audio time compression we can use memory IC's that have separate input and output terminals. The time needed to write or read a bit in one of these memories is very short, of the order of 150 ns overall, so that it is possible to interleave reading and writing actions. There is no need to clear the memory at any time, because the action of writing a bit replaces whatever was in that unit of memory previously.

The use of memory in this way therefore creates a signal which can also have 'horizontal' synchronizing pulses (at more frequent intervals) added to it. By now this is to all intents and purposes a video signal which can be recorded on a video recorder of the rotating head type. The standard video recorder modulation system of NRZ (non-return to zero, meaning that the 0's and 1's are recorded unchanged) can be used along with the circuitry that is normal to every videocassette recorder. The recorded signal can be recovered and the processing reversed so that the digital signal is taken at a steady rate from memory while it is intermittently loaded from the tape circuits into the same memory.

The recording system for the videocassette recorder has been referred to as the NRZ type, a topic that has been referred to earlier. This system is not well suited to a signal that consists of a long stream of the same bits. It is difficult to separate out exactly how many bits are in the stream, since such a stream would be recovered from the tape as a long pulse with sloping sides (Figure 5.4). The use of bits in this form, usually with one direction of

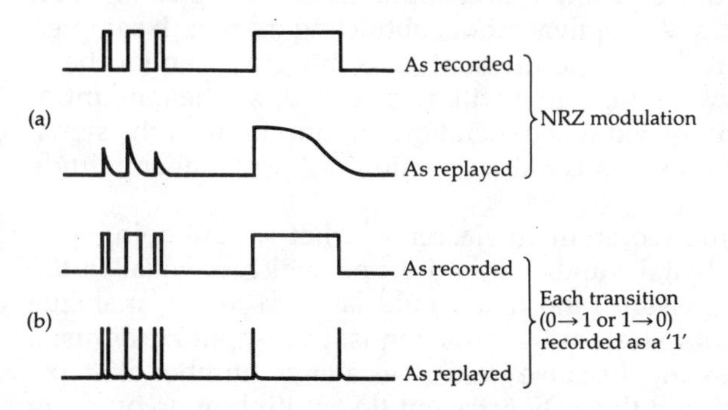

Figure 5.4 Reading problems. If a waveform contains too many grouped 1's in particular, it can cause trouble because of the inevitable integration (a). This is particularly difficult on the NRZ type of signal. If the recording method uses a 1 to signal each transition from 0 to 1 or 1 to 0 (b), then the problem is greatly reduced.

magnetization to represent 1 and the opposite direction to represent 0, is the NRZ, non-return to zero, system. It is not, as we have commented, well suited to audio coded signals, but the conversion of an audio digital signal into a frequency-modulated form of video signal allows this type of modulation to be used.

This avoids the use of the rather less simple methods that are employed for computer disc recording, and makes the recording of digital audio identical with the recording of video. As we shall see later, though, if we do not use a pseudo-video type of signal we still have to tackle the problem of having sequences of too many 1's or 0's, and this is done by altering the way that the numbers are carried in digital code, departing from the simple binary code system.

Consoles and editing

Studios that handle audio signals in digital form need to be able to carry out all the normal studio actions of fading, mixing and cutting on the digital signals, since it would be completely counter productive to have to convert back to analogue form for these actions. The splicing of tape is one action which greatly favours the use of stationary head recorders, and though machines exist to splice tape from rotary head equipment, the greater simplicity of fixed head methods can be attractive.

No such simplicity exists for fading and mixing, and a completely digital studio console is a very formidable piece of equipment. Whereas the reduction of signal amplitude for an analogue signal can be carried out by using a potentiometer, on a digital signal it has to be done by the equivalent of subtracting from each sampled digital number. The amount that is subtracted cannot be a constant, otherwise the signal will be distorted, so the amount of signal that is subtracted will depend on the amplitude of the signal – in other words it is a constant fraction of the signal amplitude at any point.

This makes the requirement clearer – what we are doing is to multiply each digital number by a fraction which is equal to the potentiometer division ratio that would be used for an analogue signal. Now multiplication by a fraction is, in computing terms, a slow process, mainly because it involves a large number of steps. The only way round this is to carry out the multiplication by using hardware; circuitry rather than operations on numbers. Small studios tend therefore to do all this type of action before the signal is converted to digital form.

Nevertheless, the completely digital studio can be created and several are in use. In the UK, the internationally respected audio engineers Neve produce all-digital consoles that are designed to be used in the production of CDs, and it cannot be long before totally digital equipment is almost universal in studios, particularly since the steadily decreasing prices of devices like memory chips have worked their way through the manufacturing system. The period in which the price of a computer for a given task fell from about £15,000 to nearer £500 has been the development period for digital sound, and we can now start to reap the benefits of the costly pioneering work.

Owners of Sony Video-8 recorders can make use of the 8mm video tapes for recording and replaying of pulse-code modulated audio signals. This can be done either along with video signals or, in another mode, as a sound-only recorder. In this latter mode, a playing time of as much as 18 hours is possible. The methods that are used are of interest in showing how digital techniques have improved in the few years since the first compact discs appeared.

The A-D converter in the Video-8 machines is of 10-bit resolution, giving a quantization to 1024 levels. When the sound is being recorded along with video signals, this 10-bit signal is reduced to 8 bits and the signals are stored into memory to be released at intervals so that the signals can be recorded at about seven times their normal rate, corresponding to about 2 Mbits per second. This audio track is recorded on the first part of each head sweep, so that the signals have to be switched at the head amplifiers. When the recorder is being used for sound only the whole of the tape can be used for the compressed audio signals.

The signals contain parity bits and are, like the signals of CD, scattered in such a way as to ensure that a dropout will affect only a small part of any signal. The analogue stages also contain a compressor for recording and an expander for replay, so that the signal range from a system of only 8 bits corresponds to that from an un-companded 13-bit system.

Tailpiece

The following description is of a scheme for an older digital recording system which used a normal videocassette machine (US standards). This gives some idea of the signal coding methods that were used, and another point of interest is that this scheme was considered at one time for domestic audio recording. As Chapter

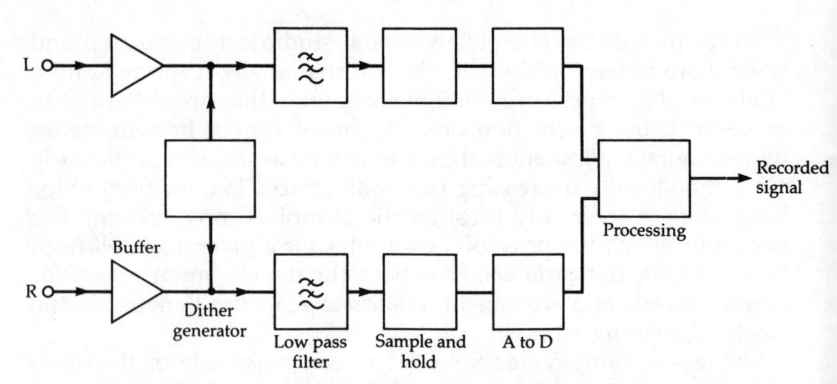

Figure 5.5 The block diagram for a digital tape recording system using pulse code modulation. The replay steps are the reverse of the recording steps.

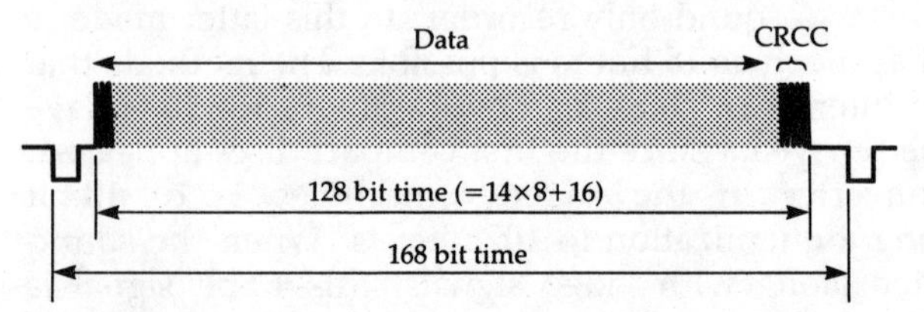

Figure 5.6 The line structure of an audio data signal created so as to simulate a colour TV waveform and allow a videocassette recorder to be used.

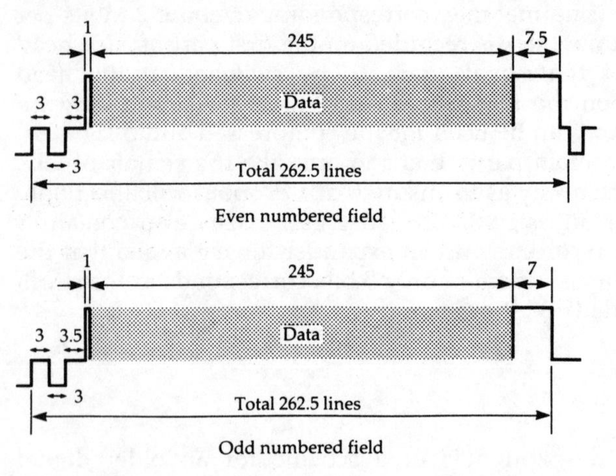

Figure 5.7 The field structure of a pseudo-video type of digital signal. This is arranged with the same type of differences between odd and even numbered fields as is used in TV signals.

7 notes, however, the final DAT system bears little resemblance to this.

The system allows for a choice of sampling rates to be used, with the maximum frequency that can be recorded depending on the sampling rate that is used. The audio signals are passed through a low-pass filter whose range is DC to 20kHz, ± 2db, so that higher frequencies do not interfere with the sampling frequency. The block diagram is shown in Figure 5.5, and you can see that the low-pass filter is followed by the sample-and-hold circuit. The output from this is converted, with a time of about 5 μs available. A sampling rate of 44.056 kHz is used.

The resulting digital signal is then changed into the form of a NTSC (the US colour TV standard) TV signal. This allows any ordinary videocassette recorder to be used for recording and replay through the video input/output socket. To start with, an eight to 14 modulation is used on the signal – this type of modulation, which gives a 14 bit signal for each eight bits of 8–4–2–1 number, is also used for CD, as noted in the following chapter.

The signal is then arranged into 'lines' and 'fields' so as to mimic the TV signal exactly. Each line takes a time equivalent to 128 bits of signal which when the synchronizing pulse is added takes a time of 168 bits, as Figure 5.6 shows. These lines are assembled into fields of 245 lines each, which with sync pulses take the time of 262.5 lines. Figure 5.7 shows the difference between 'even' and 'odd' lines which matches the interlace specification of a TV signal.

In each frame, a data disposition signal is added, consisting of one line that contains a levelling signal of 1100 (to set data level controls), a contents identification signal, a 16-bit cyclic redundancy checking code error detecting signal, and address and control signals (intended to prevent digital dubbing). The signal levels of 0.4V for 1 and 0.1V for 0 are used, corresponding to TV pedestal level and black level respectively. A peak white signal is transmitted during flyback so as to correct the action of AGC circuits.

6 The CD system

Introduction

The CD system is one rare example of international co-operation leading to a standard that has been adopted worldwide, something that was notably lacking in the first generation of videocassette recorders. The CD system did not spring from work that was mainly directed to audio recording but rather from the effort to create video discs that would be an acceptable alternative to videocassettes. At the time when competing video disc systems first became available, there were two systems that needed mechanical contact between the disc and a pickup, and one that did not. The advantages of the Philips system that used a laser beam to avoid mechanical contact were very obvious, though none of the video disc systems was a conspicuous commercial success despite the superior quality of picture and sound as compared to videocassette recorders.

The work on video discs was carried out competitively, but by the time that the effort started to focus on the potentially more marketable idea of compact audio discs, the desirability of uniformity became compelling. Audio systems have never known total incompatibility in the sense that we have experienced with video or computer systems. If you buy an audio record or a cassette anywhere in the world, you do not expect to have to specify the type of system that you will play it on. Manufacturers realized that the audio market was very different from the video market. Whereas videocassette recorders when they appeared were performing an action that was not possible before, audio compact discs would be in competition with existing methods of sound reproduction, and it would be ridiculous to offer several competing standards.

In 1978, a convention dealing with digital disc recording was organized by some 35 major Japanese manufacturers who recommended that development work on digital discs should be channelled into twelve directions, one of which was the type proposed by Philips. The outstanding points of the Philips proposal were that the disc should use constant linear velocity recording, eight-to-fourteen modulation, and a new system of error correction called Cross-Interleave Reed-Solomon code (CIRC). The use of constant velocity recording means that no matter whether the inner or the outer tracks are being scanned, the rate of digital information should be the same, and Philips proposed to do this by using a steady rate of recording or replaying data and vary the speed of the disc at different distances from the centre. You can see the effect of this for yourself if you play a track from the start of a CD (the inside tracks) and then switch to a track at the end (outside). The reduction in rotation speed when you make the change is very noticeable, and this use of constant digital rate makes for much simpler processing.

By 1980, Sony and Philips had decided to pool their respective expertise, using the disc modulation methods developed by Philips along with signal-processing systems developed by Sony. The use of a new error-correcting system along with higher packing density made this system so superior to all other proposals that companies flocked to take out licences for the system. All opposing schemes died off, and the CD system emerged as a rare example of what can be achieved by co-operation in what is normally an intensely competitive market.

The optical system

The CD recording method makes use of optical recording, using a beam of light from a miniature semiconductor laser. Such a beam is of low power, a matter of milliwatts, but the focus of the beam can be to a very small point so that low-melting point materials like plastics can be vaporised by a focused beam. Turning the recording beam on to a place on a plastic disc for a fraction of a microsecond will therefore vaporise the material to leave a tiny crater or pit, about 0.6μm in diameter – for comparison, a human hair is around 50 μm in diameter. The depth of the pits is also very small, of the order of 0.1 μm. If no beam strikes the disc, then no pit is formed, so that we have here a system that can digitally code pulses into the form of pit or no-pit. These pits on the master disc

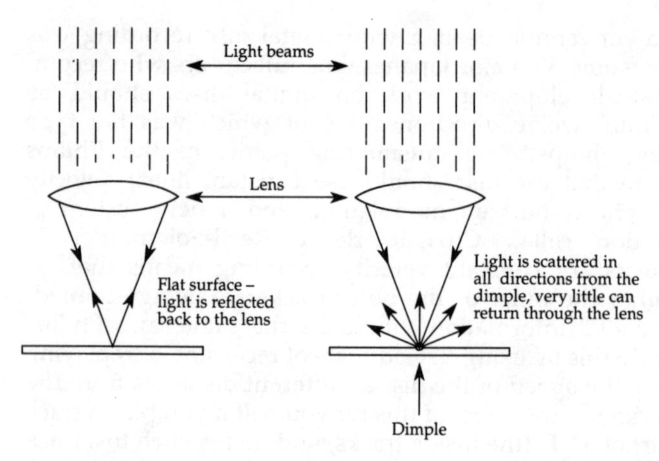

Figure 6.1 How light is reflected back from a flat surface through a lens, but scattered from a dimple. Very little of the scattered light will reach the lens.

are converted to dimples of the same scale on the copies. The pits/dimples are of such a small size that the tracks of the CD can be much closer – about 60 CD tracks take up the same width as one LP track.

Reading a set of dimples on a disc also makes use of a semiconductor laser, but of much lower power since it need not vaporise material. The reading beam will be reflected from the disc where no dimple exists, but scattered where there is a dimple, Figure 6.1. By using an optical system that allows the light to travel in both directions to and from the disc surface (Figure 6.2), it is possible to focus a reflected beam on to a photodiode, and pick up a signal when the beam is reflected from the disc, with no signal when the beam falls on to a pit. The output from this diode is the digital signal that will be amplified and then processed into an audio signal. Only light from a laser source can fulfil the requirements of being perfectly monochromatic (one single frequency) and coherent (no breaks in the wave-train) so as to to permit focusing to such a fine spot.

The CD system uses a beam that is focused at quite a large angle, and with a transparent coating over the disc surface which also focuses the beam as well as protecting the recorded pits, Figure 6.3. Though the diameter of the beam at the pit is around 0.5 μm, the diameter at the surface of the disc, the transparent coating, is about 1mm. This means that dust particles and hairs on the surface

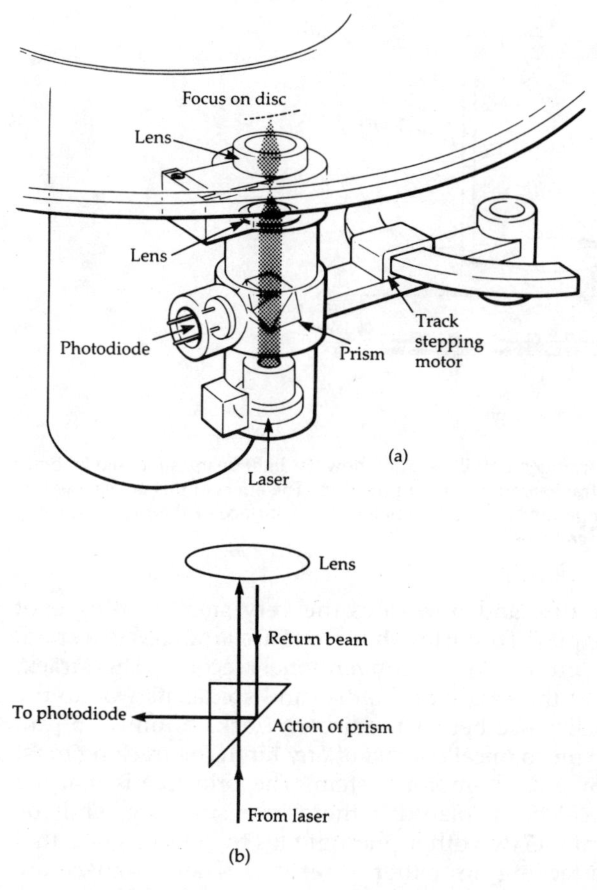

Figure 6.2 (a) The optical path of light beams in the CD player. The optical components are underneath the disc, and the purpose of the prism, shown more clearly in (b) is to split the returning light so that some of it is reflected to the photocell.

of the disc have very little effect on the beam, which passes on each side of them – unless your dust particles are a millimetre across! This is just one way in which the CD system establishes its very considerable immunity to dust and scratching on the disc surface, the other being the remarkable error detection and correction system.

Given, then, that the basic record and replay system consists of using a finely focused beam of laser light, how are the pits

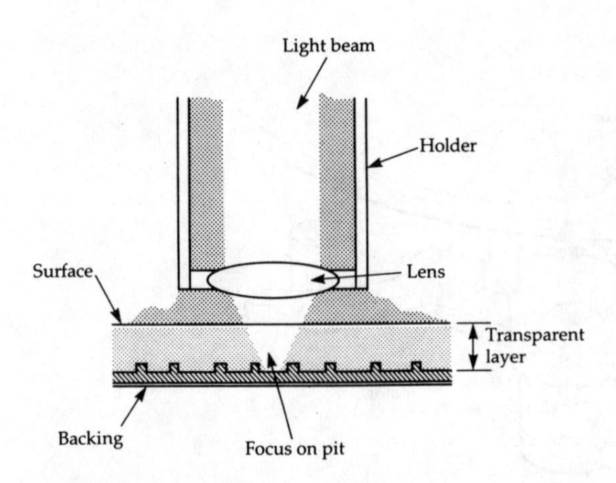

Figure 6.3 The lens arrangement, illustrating how the light beam is focused by both the lens and by the transparent coating of the disc. The effect of this is to allow the beam to have a comparatively large diameter at the surface of the disc, reducing the effect of dirt or scratches.

arranged on the disc and how does the very small reading spot keep track of the pits? To start with the pits are arranged in a spiral track, like the groove of a conventional record. This track, however, starts at the inside of the disc and spirals its way to the outside, with a distance between adjacent tracks of only 1.6 μm. Since there is no mechanical contact of any kind, the tracking must be carried out by a servo-motor system. The principle is that the returned light will be displaced if there is mistracking, and the detector unit contains two other photodiodes to detect signals that arise from mistracking in either direction. These signals are obtained by splitting the main laser beam to form two side-beams, neither of which should ever be deflected by the dimples on the disc when the tracking is correct. These side beams are directed to the photodiodes. The signals from these diodes are used to control the servo-motors in a feedback loop system, so that the corrections are being made continually while the disc is being played. In addition, the innermost track carries the position of the various music tracks in the usual digitally coded form. Moving the scanner system radially across the tracks will result in signals from the photodiode unit, and with these signals directed to a counter unit, the number of tracks from the inner starting track can be counted. This therefore allows the scanner unit to be placed on any one of the possible 41250 tracks.

Positioning on to a track is a comparatively slow operation because part of each track will be read on the way before the count number is increased. In addition, the motor that spins the disc is also servo controlled so that the speed of reading the dimples is constant. Since the inner track has a diameter of 50mm and the maximum allowable outer track can have a diameter of 116mm, the number of dimples per track will also vary in this same proportion, about 2.32. The disc rotation at the outer edge is therefore slower by this same factor when compared to the rotational speed at the inner track. The rotational speeds range from about 200 rpm on the inner track to about 500 rpm on the outer track – the actual values depend on the servo control settings and do not have to be absolute in the way that the old 33 1/3 speed had to be absolute. The only criterion of reading speed on a CD is that the pits are being read at a correct rate, and that rate is determined by a master clock frequency. This corresponds to a reading speed of about 1.2 to 1.4 metres of track per second.

The next problem is of recording and replaying two channels, because the type of reading and writing system that has been described does not exactly lend itself well to twin-channel use with two tracks being read by two independent scanners. This mechanical and optical impossibility is avoided by recording the two-channel information on one track, by recording samples from the channels alternately. This is made easier by the storage of data while it is being converted into frame form for recording, because if you have one memory unit containing L-channel data and one containing R-channel data, they can be read alternately into a third memory ready for making up a complete frame. There is no need to have the samples taken in a phased way to preserve a time interval between the samples on different channels. As the computer jargon puts it, the alternation does not need to take place in real time.

We now have to deal with the most difficult parts of the whole CD system. These are the modulation method, the error-detection and correction, and the way that the signals are assembled into frames for recording and replay. The exact details of each of these processes are released only to licensees who buy into the system. This is done on a flat-rate basis – the money is shared out between Philips and Sony, and the licensee receives a manual called the 'Red book' which precisely defines the standards and the methods of maintaining them. Part of the agreement, as you might expect, is that the confidentiality of the information is maintained, so that what follows is an outline only of methods. This is all that is needed, however, unless you intend to design and manufacture CD equipment, because for servicing work you do not need to know the circuitry of an IC or the content of a memory in order to check that it is working correctly.

The error-detection and correction system is called CIRC – Cross-Interleave Reed-Solomon Code. This code is particularly suitable for the correction of what are termed 'long burst' errors, meaning the type of error in which a considerable amount of the signal has been corrupted. This is the type of error that can be caused by scratches, and the code used for CDs can handle errors of up to 4000 bits, corresponding to a 2.5mm fault on the disc lying along the track length. At the same time, a coding that required too many redundant bits to be transmitted would cramp the expansion possibilities of the system, so the CIRC uses only one additional bit added to each three of data – this is described as an efficiency of 75%. Small errors can be detected and remedied, and large errors are dealt with by synthesizing the probable waveform on the basis of what is available in the region of the error.

The whole system depends heavily on a frame structure which, though not of the same form as a TV signal frame, allows for the signal to carry its own synchronization pattern. This makes the recording self-clocking, so that all the information needed to read

Data	12 × 16 bits (6 sets per channel)	= 24 × 8 bits		
Error Correction	4 × 16 bits	= 8 × 8 bits	33 × 8 bits	
Control/ Display		= 1 × 8 bits		(a)
		264 bits per frame		

Total data as above	33 × 14 bits	= 462	
Synchronization		24	(b)
Redundant bits for data and sync	3 × 34	= 102	
		588 recorded bits per frame	

Figure 6.4 (a) The data grouping before EFM . (b) The data grouping after EFM, with synchronization and redundancy bits added to a total of 588 bits per frame. The frame is the unit for error checking, synchronization, and control/display.

the signal correctly is contained on the disc – there is no need, for example, for each player to maintain a crystal-controlled clock pulse generator working at a fixed frequency. In addition, the presence of a frame structure allows the signal to be recorded and replayed by a rotating head type of tape recorder.

Each frame of data contains a synchronization pattern of 24 bits, along with 12 units of data (using 16-bit 'words' of data), 4 error-correcting words (each 16 bits) and 8 bits of control and display code. Of the 12 data words, six will be left-channel and six right-channel, and the use of a set of 12 in the frame allows expansion to four channels (3 words each) if needed later. Though the actual data words are of 16 bits each, these are split into 8-bit units for the purposes of assembly into a frame. The content of the frames, before modulation and excluding synchronization, is therefore as shown in Figure 6.4(a).

EFM modulation system

The word 'modulation' is used in several senses as far as digital recording systems are concerned, but the meaning here is of the method that is used to code the 1's and 0's of a digital number into 1's and 0's for recording purposes. The principle, as mentioned earlier, is to use a modulation system that will prevent long runs of either 1's or 0's from appearing in the final set of bits that will burn in the pits on the disc, or be read from a recorded disc, and calls for each digital number to be encoded in a way that is quite unlike binary code. The system that has been chosen is EFM, meaning eight-to-fourteen modulation, in which each set of eight bits is coded on to a set of 14 bits for the purposes of recording. The number carried by the eight bits is coded as a pattern on the 14 bits instead, and there is no straightforward mathematical relationship between the 14-bit coded version and the original 8-bit number.

The purpose of EFM is to ensure that each set of eight bits can be written and read with minimum possibility of error. The code is arranged, for example, so that changes of signal never take place closer than 3 recorded bits apart. This cannot be ensured if the unchanged 8 or 16-bit signals are used, and the purpose is to minimize errors caused by the size of the beam which might overlap two dimples and read them as one. The 3-bit minimum greatly eases the requirements for perfect tracking and focus. At the same time, the code allows no more than 11 bits between changes, so avoiding the problems of having long runs of 1's or 0's. In addition, three redundant bits are added to each fourteen, and these can be made 0 or 1 so as to break up long strings of 0's or 1's that might exist when two blocks of 14 bits were placed together.

The idea of using a code which is better adapted for reading and writing is not new. The Gray code has for a long time been used as an alternative to the 8-4-2-1 type of binary code, because in a Gray code a small change of number causes only a small change of code. The change, for example from 7 to 8 which in four-bit binary is the change from 0111 to 1000 would in Gray code be a change from 0100 to 1100 – note that only one digit changes. EFM uses the same principles, and the conversion can be carried out by a small piece of fixed memory (ROM) in which using an eight-bit number as an address will give data output on 14 lines. The receiver will use this in reverse, feeding 14 lines of address into a ROM to get eight bits of number out. If you want to find the exact nature of the code, buy the Red Book!

The use of EFM makes the frame considerably larger than you would expect from its content of 33 8-bit units. For each of these units we now have to substitute 14 bits of EFM signal, so that we have 33 x 14 bits of signal. In addition, there will be 3 additional bits for each 14-bit set, and we also have to add the 24 bits of synchronization and another three redundant bits to break up any pattern of excessive 1's or 0's here. The result is that a complete frame as recorded needs 588 bits, as detailed in Figure 6.4. All of this, remember, is the coded version of 12 words, six per channel, corresponding to 6 samples taken at 44.1 kHz, and so representing about 136 μs of signal in each channel.

Error correction

The EFM system is by itself a considerable safeguard against error, but the CD system needs more than this to allow, as we have seen earlier, for scratches on the disc surface that cause long error sequences. The main error-correction is therefore done by the CIRC coding and decoding, and one strand of this system is the principle of interleaving. Before we try to unravel what goes in the the CIRC system, then, we need to look at interleaving and why it is carried out.

The errors in a digital recording and replay system can be of two main types, random errors and burst (or block) errors. Random errors, as the name suggests, are errors in a few bits scattered around the disc at random and, because they are random, they can be dealt with by relatively simple methods. The randomness also implies that in a frame of 588 bits, a bit that was in error might not be a data bit, and the error could in any case be corrected reasonably easily. Even if it were not, the use of EFM means that an error in one bit does not have a serious effect on the data.

A block error is quite a different beast, and is an error that involves a large number of consecutive bits. Such an error could be caused by a bad scratch in a disc or a major dropout on tape, and its correction is very much more difficult than the correction of a random error. Now if the bits that make up a set were not actually placed in sequence on a disc, then block errors would have much less effect. If, for example, a set of 24 data units (bytes) of 8-bits each that belonged together were actually recorded so as to be on eight different frames, then all but very large block errors would have the effect of a random error, affecting only one or two of the byte units. This type of shifting is called interleaving, and it is the most effective way of dealing with large block errors. The error-detection and correction stages are placed between the channel alternation stage and the step at which control and display signals are added prior to EFM encoding.

The CIRC method uses the Reed-Solomon system of parity coding along with interleaving to make the recorded code of a rather different form and different sequence as compared to the original code. Two Reed-Solomon coders are used, each of which adds four parity 8-bit units to the code for a number of 8-bit units. The parity system that is used is a very complicated one, unlike the simple single-bit parity that was considered earlier, and it allows an error to be located and signalled. The CD system uses two different Reed-Solomon stages, one dealing with 24 bytes (8-bit units) and the other dealing with 28 bytes (the data bytes plus parity bytes from the first one), so that one frame is processed at a time. In addition, by placing time delays in the form of serial registers between the coders, the interleaving of bytes from one frame to another can be achieved. The Reed-Solomon coding leaves the signal consisting of blocks that consist of correction code(1), data(1), data (2) and correction code(2), and these four parts are interleaved. For example, a recorded 32-bit signal might consist of the first correction code from one block, the first data byte of the adjacent block, the second data byte of the fourth block, and the second correction code from the eighth data block. These are assembled together, and a cyclic redundancy check number can be added.

At the decoder, the whole sequence is performed in reverse. This time, however, there may be errors present. We can detect early on whether the recorded 'scrambled' blocks contain errors and these can be corrected as far as possible. When the correct parts of a block (error code, data, data, error code) have been put together, then each word can be checked for errors, and these corrected as far as possible. Finally, the data is stripped of all added codes, and will either be error-free or will have caused the activation of some method (like interpolation) of correcting or concealing gross errors.

Production methods

A compact disc starts as a glass plate which is ground and polished to 'optical' flatness – meaning that the surface contains no deformities that can be detected by a light beam. If a beam of laser light is used to examine a plate like this, the effect of reflected light from the glass will be to add to or subtract from the incident light, forming an interference pattern of bright and dark rings. If these rings are perfectly circular then the glass plate is perfectly flat so that this can be the basis of an inspection system that can be automated.

The glass plate is then coated with photoresist, using the types of photoresist that have been developed for the production of ICs. These are capable of being 'printed' with very much finer detail that is possible using the type of photoresist which is used for PCB construction, and the thickness and uniformity of composition of the resist must both be very carefully controlled.

The image is then produced by treating the glass plate as a compact disc and writing the digital information on to the photoresist with a laser beam. Once the photoresist has been processed, the pattern of pits will develop and this comprises the glass master. The surface of this disc is then silvered so that the pits are protected, and a thicker layer of nickel is then plated over the surface. This layer can be peeled off the glass and is the first metal master. The metal master is used to make ten mother plates, each of which will be used to prepare the stamper plates.

Once the stamper plates have been prepared, mass production of the CDs can start. The familiar plastic discs are made by injection moulding (the word 'stamper' is taken from the corresponding stage in the production of black vinyl discs) and the recorded surface is coated with aluminium, using vacuum vaporization.

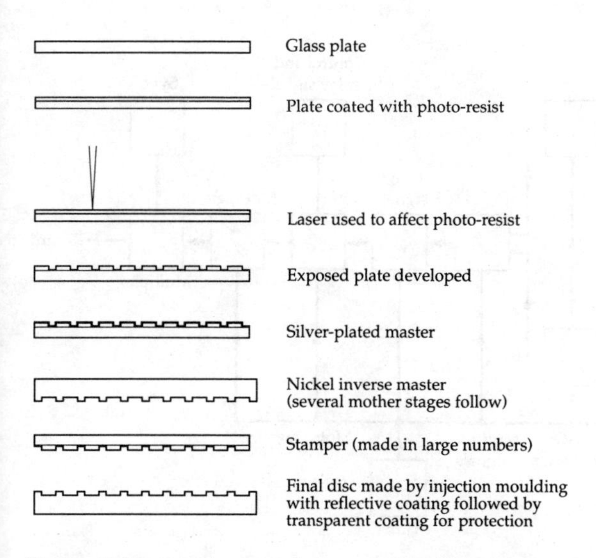

Figure 6.5 Steps in the manufacture of a CD. The stampers are used in the injection moulding process, and the plastic discs have to be aluminised to reflect light, and then treated with their transparent coating which protects the dimples and helps to focus the laser light.

Following this, the aluminium is protected by a transparent plastic which forms part of the optical path for the reader and which can support the disc label. The system is illustrated in outline in Figure 6.5.

The end result

All of this encoding and decoding is possible mainly because we can work with stored signals, and such intensive manipulation is tolerable only because the signals are in the form of digital code. Certainly any efforts to correct analogue signals by such elaborate methods would not be welcome and would hardly add to the fidelity of reproduction. Because we are dealing with signals that consist only of 1's and 0's, however, and which do not provide an analogue signal until they are converted, the amount of work that is done on the signals is irrelevant. This is the hardest point to accept to anyone who has been brought up in the school of thought that the less that is done to an audio signal the better it is. The whole point about digital signals is that they can be manipulated

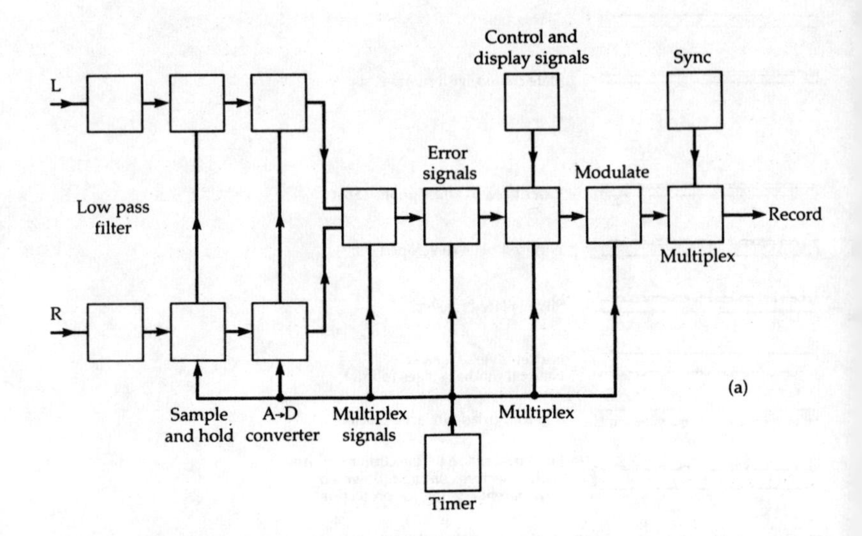

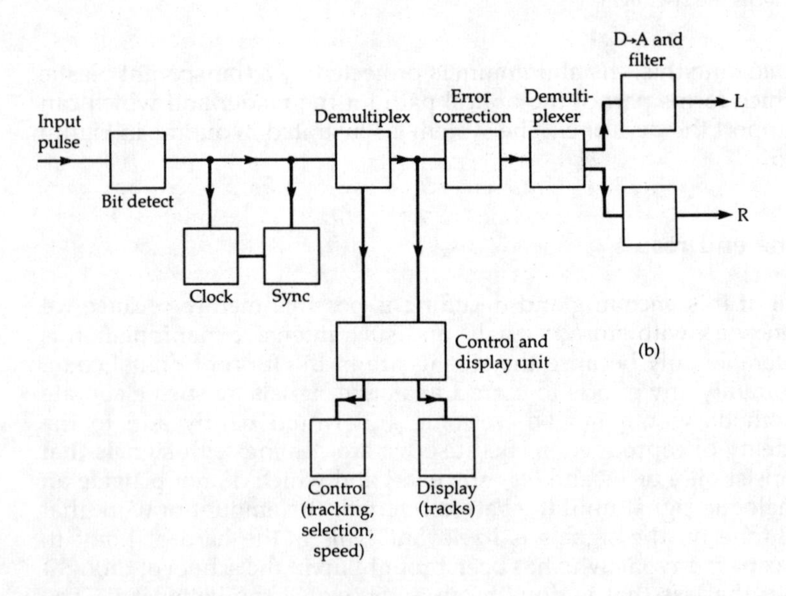

Figure 6.6 Block diagrams for CD recording (a) and replay (b).

as we please providing that the underlying number codes are not altered in some irreversible way.

The specifications for the error-correcting system are:

Max. correctable burst error length 4000 bits (2.5 mm length)

Maximum interpolable burst error length 12,300 bits (7.7 mm length). Other factors depend on the bit error rate, the ratio of the number of errors received to the number of bits total. The system aims to cope with bit error rates (BER) between 10^{-3} and 10^{-4}.

For BER = 10^{-3} interpolation rate is 1000 sample per minute.
– undetected errors less than 1 in 750 hours

For BER = 10^{-4} interpolation rate is 1 sample in 10 hours.
– undetected errors negligible.

Control bytes

For each block of data one control byte (8 bits) is added. This allows eight channels of additional information to be added, known as sub-codes P to W. This corresponds to one bit for each channel in each byte. To date, only the channels P and Q have been used. The P channel carries a selector bit which is 0 during music and lead-in, but is set to 1 at the start of a piece, allowing a simple but slow form of selection to be used. This bit is also used to indicate the end of the disc, when it is switched between 0 and 1 at a rate of 2Hz in the lead-out track.

The Q channel contains more information, including track number and timing. A channel word consists of a total of 98 bits, and since there is one bit in each control byte for a particular channel the complete channel word is read in each 98 blocks. This word includes codes which can distinguish between four different uses of the audio signals:

2 audio channels, no pre-emphasis
4 audio channels, no pre-emphasis
2 audio channels, pre-emphasis at 50μs and 15μs
4 audio channels, pre-emphasis at 50μs and 15μs

Another four bits are used to indicate the mode number for the Q channel. Three such modes are defined, but only mode 1 is of interest to users of CDs. When mode 1 is in use, the data for the disc is contained in 72 bits which carry information in nine 8-bit sections. These are:

1. TNO – the track number byte which holds two digits in BCD code. The lead-in track is 00 and the others are numbered in sequence.
2. TOC (Table of Contents) information on track numbers and times, three bytes
3. A Frame number and running time within a track, three more bytes.
4. A zero byte.

The allocation of these control channels allows considerable flexibility in the development of the CD format, so that players in the future could make use of features which have to date not been thought of. Some of the additional channels are used on the CDs which are employed as large-capacity storage for computers.

Summary

The proof of the efficacy of the whole system lies in the audio performance. The bare facts are impressive enough, with frequency range of 20Hz to 20 kHz, (within 0.3 db) and more than 90db dynamic range, signal to noise ratio and channel separation figures, with total harmonic distortion (including noise) less than 0.005%. A specification of that order with analogue recording equipment would be difficult, to say the least, and one of the problems of digital recording is to ensure that the analogue equipment which is used in the signal processing is good enough to match up to the digital portion.

Added to the audio performance, however, we have the convenience of being able to treat the digital signals as we treat any other digital signals. The inclusion of control and display data means that the number of items on a recording can be displayed, and we can select the order in which they are played, repeating items if we wish. Even more impressive (and very useful for music teachers!) is the ability to move from track to track, allowing a few notes to be repeated or skipped as required. Though some users treat a CD recording as they used to treat the old black-plastic disc, playing it from beginning to end, many have come to appreciate the usefulness of being able to select items, and this may be something that recording companies have to take notice of in the future. The old concert-hall principle that you can make people hear modern music by placing it between two Mozart works will not apply to CD, and the current bonanza of re-recording old master tapes (often with very high noise levels) on to CD cannot last much longer.

Every real improvement in audio reproduction has brought with it an increase in the level of acceptability. We find the reproduced sound quality of early LP's now almost as poor compared to our newly raised standards as the reproduction from 78's appeared in the early days of LP's. The quality that can be obtained from a CD is not always reflected in samples I have heard, where the noise from the original master tapes and the results of editing are only too obvious. Music lovers are usually prepared to sacrifice audio quality for musical quality – we will listen to Caruso or McCormack with a total suspension of criticism of the reproduction – but only to a point. Nothing less than full digital mastering coupled with excellent microphone technique and very much less 'sound engineering' used on the master tapes is now called for, and this will mean a hectic future for orchestras as the concert classics are recorded anew with the equipment of the new generation.

In case it all sounds too perfect, there can still be problems in production. I have a CD of archive recordings of Maria Callas in which many tracks are very poor because of either overmodulation or bad tracking caused by poor CD production techniques. The symptom is a loud clicking at about the rate of revolution of the disc, or strange metallic noises on parts of the music. Discs of this quality should never have left the factory (in Germany), and if you do come across a faulty disc you should return it at the first possible opportunity so that the factory can be aware of the problems – I shall keep mine as a dreadful warning that there is no such thing as perfection.

Bitstream tailpiece

The use of bitstream technology was mentioned in Chapter 4. Its advantages as far as CD reproduction is concerned are fourfold:

1. The conversion from digital to analogue is now carried out in a digital way, with no need to generate 16 very precise values of current.
2. The linearity of small signals is greatly improved because of the elimination of cross-over distortion caused by current inaccuracies.
3. The glitches that were caused in conventional D-A converters when bit values changed at slightly different times are eliminated.

4. The zero level (I/2), known as digital silence, is very much more stable because it no longer depends on accuracy of current division, only on the generation of a 101010.. pattern from the bitstream converter.

The Philips system uses an oversampling of 256 times to generate a signal stream at 11.2896 MHz. A dither signal at 176 kHz is added to reduce the formation of patterns in the noise shaper, and this is addition of dither is the equivalent of adding another bit to the 16-bit signal. In the conversion process, the quantization noise which is generated when a 16-bit word is reduced to a stream of single bits is all in the higher frequency band and is removed by a simple low-pass filter.

The simple system described in Chapter 4 shows the action of a first-order noise-shaper, with one stage of integration only and acting on one input number only. The Philips system uses a second-order noise shaper, working with two delays and adding to two waiting input numbers. This provides an even higher standard of performance, typically of distortion at 1 kHz 92 db down, small signal distortion 94 db down and signal-to-noise ratio of 96 db, with linearity of ±0.5 db.

The IC which is used for bitstream conversion can be used in a differential mode, with inverted data being fed to another unit and the output combined with inversion. This has the effect of adding the in-phase signals but subtracting noise signals, leading to a further improvement in signal-to-noise ratio.

The MASH system

The principle of using two-level outputs has been taken up by several other manufacturers, and Technics refer to their system as MASH, an abbreviation of 'multi-stage noise shaping', first developed by NTT Inc. and developed by Matsushita Electric. The MASH system uses a third order noise-shaping circuit, previously thought to be susceptible to oscillation, but designed so as to ensure stability with a very precise conversion of digital to analogue conversion in the audio band. A crystal oscillator is used to generate the delay time, so ensuring very high accuracy along with excellent temperature and time stability.

The sampling is effectively at 768 times the CD sampling frequency, so that the output from the three noise shapers is at 33.869 MHz. The output signal is 3-bit (7 levels) rather than 1-bit, and the seven levels of this 3-bit code are converted into pulse width variations. This allows the final conversion to analogue to make use of simple low-pass filters.

Another system, PEM (Pulse Edge Modulation) is being used by JVC, with sampling at 384 times the normal rate and four noise shapers to provide a bitstream at 16.934 MHz.

A rather different approach has been taken by the Denon DAT players, using the name of LAMBDA SLC, Ladder-form Multiple Bias Digital to Analogue Super Linear Conversion. This method uses a 20-bit D-A conversion in duplicate, with the outputs combined so as to reduce errors.

7 Consumer digital systems

Disc recording, as far as the consumer has been concerned, has always been read-only apart from a few machines that were marketed in the 30's using aluminium discs. Tape recording, by contrast, has always been read/write, allowing the user the choice of replaying commercially-made tapes or of making recordings from broadcast material. In theory, the tape user respects copyright and never copies discs, but the coming of the cassette unit ensured that disc copying would be widespread, if only to allow music on disc to be played in a car. The copyright laws were never intended to cope with copying on such a scale, and in practice were used only against large-scale copying for gain (pirated tapes) where breach of copyright could be proved.

Problems encountered with analogue tape

Tape for consumer uses, whether open-reel or cassette, was initially analogue, and as a consumer recording medium, using in cassette recorders almost exclusively, it presents many severe technical problems. These are:

1. The inherent non-linearity of the tape recording process, dictated by the shape of the magnetic hysteresis curve.
2. The high noise level of tape.
3. The poor 'headroom', meaning that severe distortion results when the tape is overloaded.
4. The poor frequency response of slow-moving tape.

All of these factors made the cassette totally unsuited to quality recording when the medium was first introduced; they also ensured that copies of tapes were of poor quality. Since the introduction of the cassette (more correctly, the compact cassette), improvements in tape coatings, record/replay heads and, particularly, circuit methods such as Dolby and dbx have made the use of analogue cassettes for good quality recording achievable, and sales of 190 million cassette players each year worldwide confirm how thoroughly this medium has achieved success – and how difficult it will be to replace.

Considering the unsatisfactory nature of analogue magnetic tape recording, and particularly the narrow-tape formats like cassettes, it is surprising that the first consumer application of digital recording was to disc rather than to tape. It is all the more surprising when we consider that videocassette recording became available at about the time that studios were switching to digital tape recorders, so that the appearance of an audio digital tape system at that time might have been expected on technical grounds at least. Several audio enthusiasts bought one model of Sanyo videotape machine on the grounds that it permitted switching to audio recording, using the principles outlined in Chapter 5. This allowed three hours of really high-class recording to be made on a cheap videotape, as opposed to an hour or so on an open-reel tape recorder using a £15 tape spool.

The reasons for the emergence of CD rather than digital tape as the first consumer digital reproduction system lies more in established practice than in technology. Disc has always been the preferred medium for distribution of high-quality sound recordings. Discs have historically been of higher audio standard and much easier to mass produce than tapes, even cassette tapes. Though sales of music on cassette have risen steadily in the past few years to overtake sales of disc, most of these cassettes are not of an audio quality that would appeal to the potential user of CD. As far as buying recorded high-quality music was concerned, the disc was the medium to use. This was the reason for concentrating on CD; and in any case the technical problems had already been researched during the development work on videodiscs. When analogue tape systems are used, the worst effects are mitigated by the use of noise-reduction systems such as Dolby-B, and results equivalent to LP quality can be obtained by using methods such as Dolby-C or dbx. An important advantage of tape is that the popular C90 size of cassette offers 45 minutes of playing per side, double the time of the LP.

The tape cassette, in addition, is still the most convenient medium to carry around, to use in the car, and certainly the only one that can be recorded as well as being played. The videodisc never made much serious impact because the videocassette recorder was used primarily as a way of time-shifting transmitted programs, the ideal solution to the usual situation that the only two TV programs you want to watch in a week are always at the same time on the same evening. A player that offered replay-only of films at high prices was not particularly appealing – few films are worth seeing more than once, and hiring from the local video shop is usually cheaper and simpler than waiting for a film to become available on disc (or on cable or satellite). Problems like that don't arise in audio – quite apart from anything else, not many listeners can receive Radio 3 on FM with *really good* signal strength and the local radio that uses up all the money that might otherwise be used to improve Radio 3 is not exactly geared to music. In any case, if you were given a high-quality tape system, what would be worth recording – unless you already had a CD player?

The problem of copying

The factor that has, more than any other, delayed the appearance of digital tape has been the fear of undercutting the profitable market for discs. The manufacture of CD's requires enormous investment, and unless this can be recovered no manufacturer is likely to contemplate setting up a plant to make discs. If each disc can be transcribed to tape with no loss of quality, and more seriously, if that tape can be used to make another and so on, all with the quality of the original preserved, then the risk is obvious. Unlike analogue tape recording, in which each copy is noticeably inferior to the recording from which it was made, digital tape offers the opportunity to make copies that are of equal quality, even if there have been hundreds of recording stages between master and copy. The sale of digital tape equipment in many countries has for some time been resisted until a method of limiting the extent to which CDs could be copied had been worked out. Such an agreement has, in the UK at least, been achieved by the incorporation of circuits into digital tape machines which allow a copy to be made of a CD, but do not permit that tape to be used to make a second generation copy. The system is described as the SCMS – Serial Copy Management System. This is a satisfactory solution which allows you to make a copy of each of your own CDs for your own use, but deters multiple-copying for commercial gain.

At the time of writing, then, the standards for DAT have been set and domestic machines are available, though at a price level very much higher then CD players. The specifications of a

Denon machine will be examined later in this chapter. In addition we shall look at systems that are likely to provide severe competition for DAT, the Philips DCC system and the Sony Mini Disc.

The systems

The two systems of digital audio tape (DAT) that have vied for consumer attention are the stationary head systems, DASH and S-DAT and the rotary head system, R-DAT. The stationary head systems have the considerable potential advantage of allowing simple tape-cut editing, something that is not possible for the domestic user of the R-DAT system, nor needed by most users. The advantage is potential only, because if data is recorded with interleaving, simple cut-and-join editing is not possible because the signals are not in their correct order. The DASH system is intended for professional use, and for a user who does not want to use editing or multi-channel recording, the S-DAT system has no compelling advantages, and is likely to be more costly. A brief look at DASH and S-DAT is useful if only to emphasise how different it is.

The DASH system

The earlier DASH-1 recorders used ferrite heads with 8 digital tracks on 1/4″ tape or 24 tracks on 1/2″ tape. The later DASH-2 standard makes use of thin-film heads which allow double this number of tracks to be used. In addition to the audio tracks, the tape must allow for an auxiliary track for timecodes and other control signals along with cue tracks. Three tape speeds have been used, corresponding to three different sampling rates. These are 50.8 cm/s (20″/s) speed with 32kHz sampling, 70 cm/s (27.56″/s) with sampling at 44.1kHz and 76.2 cm/s (30″/s) with sampling at 48kHz. These tape speeds are, of course, considerably higher than any used in domestic tape recording, even with open-reel machines.

There are also three options for track allocation which can affect tape speed. Using DASH-F, one audio channel is recorded on one tape track and the tape speed is maximum. For DASH-M two tape tracks share a single audio channel and half the tape speed of

DASH-F. The DASH-S version uses four tape tracks for one audio track and can run at quarter the speed of the DASH-F version.

The S-DAT system

The S-DAT cassette that has appeared to date measures 86 x 55 x 10 mm, and is slightly larger than the R-DAT type. The speeds that have been used are 4.76 cm/s, identical to that for the conventional audio cassette, or the slightly slower speed of 4.37 cm/s which gives longer recordings. The tape is considerably wider than the conventional audio cassette tape, however.

The density of data on the recorded tape is 64000 bits per inch, and this is achieved by using a multitrack head. This is a far cry from the type of multitrack heads that are used in conventional audio recorders, because these digital heads are made by thin-film techniques that allow a remarkable number of tracks to be laid down on a given width of tape. This is how the problem of tape speed has been overcome, since the head gaps that are currently used on high quality cassette machines are already as small as current technology allows.

The R-DAT system

The winner of the first round of technology for consumer use, however, is the R-DAT system, using a recorded wavelength of 0.7 microns. The basis of R-DAT is a cassette which uses tape that is 3.8 mm wide to match the tape of the ordinary compact cassette. The cassette itself is 73 x 54 x 10.5 mm, slightly smaller than the established compact cassette (Figure 7.1). This tape is taken in a circular path across a guide, and scanned by two record/replay heads in the same way as is used for video recording. All similarity, however, ends there.

The original R-DAT designs were all intended to use recording that was compatible with video recording, using standard video tapes. This would have used a line and field structure that was identical to the US NTSC television signal, so that it could have been recorded on US recorders. Though there is no essential reason for objecting to this, the use of this standard would not permit the replay of such tapes using European video recorders, which use different video standards. We have already looked at the brief specification for the circuitry of such a recording system

Figure 7.1 Digital audio tape package.

in Chapter 5. The older VHS and Beta system tapes are too large for such uses, but the Sony Video-8 is a very attractive possibility, since the cassette is about the same size as a normal audio cassette. Sony Video-8 recorders and cameras allow these 8mm cassettes to be used for recording up to 18 hours of high-quality digital sound per cassette in place of their normal use for both video and sound tracks.

Since audio digital tape for the consumer demands new equipment in any case, the agreed standard has changed considerably since the early days. Full details of the standards are available only to participating manufacturers, and the agreement of all 81 participating manufacturers is needed before technical information can be divulged. What follows, therefore, is based on such information as has been released and is freely available in the form of articles and lectures.

The standards that have emerged owe very little to previous suggestions, other than the use of rotating heads. In particular, there is absolutely no attempt to simulate a video signal, and this is a particularly important difference. Apart from anything else, it divorces rotating head audio tape technology from video tape technology, and puts an end to the idea that users might buy modified video recorders, or connect digital converters, to play their audio tapes. This does not, of course, prevent the manufacturers of video recorders from incorporating rotating audio heads into their machines, as a number have already done, particularly on Video-8 machines which now form a very substantial fraction of the video camcorders being sold.

The mechanics of the system

R-DAT makes use of a 90 degree wrap of tape around its guides. This is a much smaller angle than is used in video recorders, and it makes the tape path much simpler, avoiding the elaborate tape winding step which is needed when a video tape is loaded into a machine. In addition, this smaller angle greatly reduces tape wear and lessens the likelihood of tape breakage, since audio cassettes usually come in for rougher treatment than video cassettes. It also allows for the heads to maintain contact with the tape during fast-wind at some 200 times normal speed so that indexing signals can be used to mark out one section of the tape from another.

Two revolving heads are used, so that in one revolution of the head spindle, with a 90 degree wrap, the tape is being read or written by a head for only 50% of the time. This of course makes the time for which no head is in contact with the tape also 50% of the total, unlike the video recording systems in which there is an overlap period in which both heads will be in contact with the tape. The smaller time of contact is possible because R-DAT signals make use of time-compression so that the signals are gathered up in a memory, and recorded only while a tapehead is in contact with the tape. The amount of memory that has to be used for this purpose depends on the amount of data compression that is being used, and several standards exist.

The mechanical details of the system are as follows. The head drum is of 30 mm diameter, and its speed of rotation is 2000 rpm. The speed of the tape itself is low, with three rates of 4.075 mm/s (half-speed), 8.150 mm/s (normal speed) and 12.225 mm/s (wide-track) standardized. The angle of the track laid down by the recording heads is 6°22′59.5″, giving a track length of 23.501 mm in normal modes. In wide track mode, the angle is slightly greater, giving a track length of 23.471 mm. These half-speed and wide track modes are optional extras that allow for either extra long playing time at the expense of quality or extra quality at the expense of playing time respectively. The width (pitch) of each recorded track is 13.591 μm in normal mode and 20.41 μm in wide track mode. The head gap is at an azimuth angle of plus or minus 20°, alternating from one track to the next and recording over a width of 2.613mm. These mechanical layouts are summarized in the drawing of Figure 7.2. The term 'frame' is still used and defined as one pair of tracks, one in each azimuth direction.

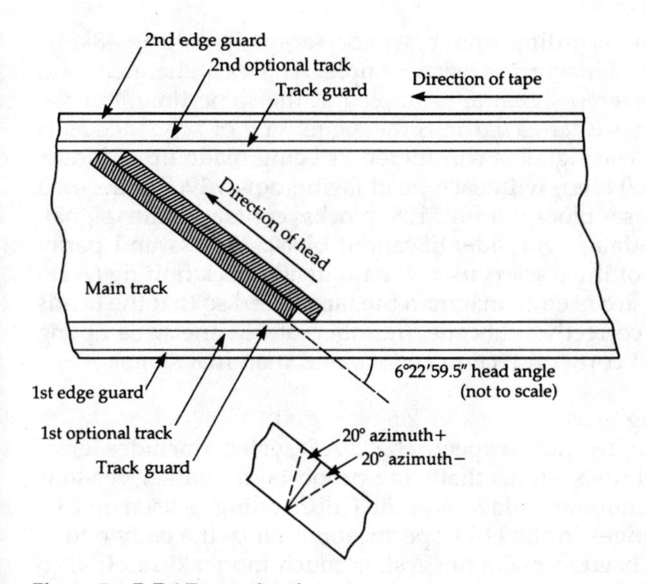

Figure 7.2 R-DAT tape detail

The specification for R-DAT allows for four possible sampling rates, including 32kHz, 44.1kHz and 48kHz. The manufacturer of any piece of equipment can select a sampling rate, and it is likely that the 44.1kHz rate, identical to the rate used for CD, will be confined to play-only machines, so as to avoid the possibility of easily recording digital signals directly from CD on to DAT. The other rates are available for record/play machines, and it remains to be seen what the final choice for consumer equipment in the UK will be. The standards allow for six possible modes of use, combining these sampling rates with different bit numbers according to the use of the tapes.

The reason for the multiple standards is to allow for development. DAT is already being used as an (expensive) system for the backup for computer hard discs, and there would be little point in using a system that was intended for audio for such purposes. In addition, the DAT standards are intended to govern a whole range of DAT type recorders, including professional studio recorders, and copying machines, so that the existence of six modes does not mean that DAT players will come equipped with a six position selector switch, only that the standard that is used for domestic machines will use one of these modes, as agreed by the 81 participants.

The normal recording and playback sampling rate is 48kHz, providing a 16-bit word at each sample. The two channels of a conventional stereo signal are sampled at the same time, but the signals are then interleaved into the same type of sequence as is used for CD. The signal is considered as being made up of blocks of 45 bytes (360 bits), with each head laying down 196 blocks on a track. Of these blocks, only 128 blocks contain main signal, consisting of data, sync, identification, block address and parity bits. The rest of the track is used for automatic track finding (ATF) signals which are used to maintain the tape speed so that the heads are tracking correctly, subcode (number of channels, sampling frequency and copy protection codes), and marginal signals.

Error checking

As you might by now expect, the DAT system includes error checking, and the systems that were proposed for earlier versions have been abandoned in favour of the CIRC coding as used on CD. The main changes in the DAT specifications since the early efforts have, in fact, been to make the system much more akin to CD, so that the record companies can use the same signal processing equipment for most of the chain from microphone to recording medium, with less risk of problems arising from the use of different systems.

The raw sampled data consists of the two channels, each sampling sixteen bits at a sampling rate of 48kHz, giving a bit rate of 2 × 16 x 48kbit/s, or 1536kbit/s, equal to 1.500Mbit/s. Two Reed-Solomon coders are used, following which the bits are interleaved to reduce the effect of long path errors. The redundancy that the Reed Solomon coding adds, plus the effect of the addition of sub-data, increases the rate to 2.77Mbit/s. In addition, there is track interleaving of blocks. One track in a pair will contain the even numbered blocks of the right hand channel and the odd numbered blocks of the left hand channel. The other track of a pair will contain the odd numbered blocks of the right hand channel and the even numbered blocks of the left hand channel. This additional interleaving allows for the complete loss of one track, so that data could be filled in from the neighbouring track. The bunching of the signals by memory so that they can be recorded in bursts by the heads then increases the bit rate at the recording head to around 7.5Mbit/s.

Copy protection

The problem of copying CDs has been tackled by using the agreed

Serial Copy Management System which adds copy protection codings to the subcode. This does not, of course, prevent anyone who is inclined to ignore copyright from copying the analogue signal from a CD player to the audio input of a DAT recorder. What is does do, however, is to prevent digital copying of the digital signals. The difference is important, because copying digital signals results in perfect copies, whereas the conversion to analogue and back will introduce noise and distortion in the analogue stages so that the copy is not perfect. It is also likely that the different sampling rates would cause beat signals to appear on a tape created in such a way.

This is not likely to deter a professional audio pirate, of whom the Far East (and, alas, the near West) has more than its share, but it should fulfil the need to deter amateurs from mass copying. This, after all, was the original intention of the copyright acts, which were framed to prevent commercial copying at a time when no-one foresaw the possibility of a domestic machine which could make recordings.

The first DAT machines are now available for sale, though travellers to the Far East have been able to buy both machines and tapes for some time now. The first applications for DAT in Europe were, in fact, for computer uses, and we shall probably see several enthusiasts trying to adapt these machines for audio signals. The prices of recorders are high at present, but once agreements have been reached for marketing DAT, then there is no reason why prices should not drop to about the same level as CD players. There is much less certainty about a supply of recorded tapes. Blank cassettes are already available from a number of suppliers, but no-one seems inclined to make any major commitment to releasing music on DAT. We may, in fact, see a circular argument developing here, with recorders unavailable because of the lack of tapes, and tapes unavailable because of the lack of recorders. The other question is whether there is any genuine demand for a high-cost high-quality tape system which is incompatible with all other tape systems, particularly when purchasing a video-8 camcorder at about the same price (or less, in many examples) as has been quoted for an R-DAT system will allow the use of 8mm cassettes for sound reproduction. On the face of it, the two appear to be mutually exclusive, and to many users it might seem logical to carry out all recording, audio and video, on the same 8mm cassettes.

We could also query whether the conventional analogue tape is dead, because Dolby Laboratories, whose experience with both analogue and digital signal processing is second to none, have demonstrated the new Dolby-S analogue coding system. Switching between a CD source and a Dolby-S recording made from the same CD has shown that listeners cannot reliably distinguish between the two even when very high-quality amplifiers and speakers are used. In addition, Dolby Laboratories have been working on tape system which use a form of delta modulation that allows a very considerable reduction in the bandwidth that is required for digital recording.

In addition, there are several other developments pending which might make R-DAT go the same way as domestic Beta-system video recorders. One is the announcement by Philips of the Digital Compact Cassette (DCC) (see below) which uses the conventional audio cassette which can be recorded with either conventional audio tracks or a digitally recorded signal. The use of this format means that the new blank cassettes can be used either for analogue or for digital recording, and the aim is that the digital recorders will also be able to play the analogue recordings at normal analogue quality.

The other development applies to analogue recording and is a patent for a new tape bias system called contour biasing which avoids the compromises that always attend conventional AC bias systems. This system is claimed to reduce tape-hiss and permit better recording of high frequencies to such an extent that a tape recorded without any noise-reduction systems can be as free from noise as a CD. The very considerable advantage is that this could be applied to pre-recorded analogue tapes and the benefits noticed even on simple replay systems.

One point that does seem to be clear is that the copyright acts will have to be drastically revised. We can no longer pretend that people buy tape recorders in order to make tapes of baby's first words or to play poorly-produced and badly recorded commercial cassettes, just as we cannot pretend that owners of computers do not make copies of every disc that comes their way. Until it is possible for a private owner of domestic equipment legally to make copies from any source of any material for his/her own non-commercial use, the copyright law will be held in contempt and regarded as an unenforceable and antique provision. This will result in gains only for the commercial pirates.

The DCC system

The principles of Digital Compact Cassette (DCC), a stationary head system, contain items of considerable importance for the future development of all recording systems, because they hinge on much more complex signal processing than has been used in the past for audio-only signals. When compact disc was developed, the conversion from analogue to digital was total; the analogue waveform was simply converted to a digital waveform with no attempt to alter its characteristics. This led automatically to the bandwidth requirements that have determined the standards for the CD system. This is still the standard as far as high-quality sound is concerned, and there is likely to be furious argument over the use of 'doctored' sound in the new systems.

Psychological factors

The use of 8mm video tapes for sound recording (by Sony) started the quest for reduction of bandwidth. Two of the three factors which make such reduction possible are psychological rather than technical, which is why there will be arguments on the subjective sound quality for some considerable time to come.

The first psychological argument concerns the threshold of hearing. For each frequency in a sound, there is a minimum amplitude below which the sound is not heard. This amplitude is lowest for a frequency around 2.5kHz, highest in the extremes of the spectrum (20Hz and 20kHz) and though the effect varies from one person to another (and alters with age) an average value can be taken which applies to most of the human race. CD recording takes no notice of this, using valuable digits for recording sounds that cannot (in theory at least) be heard.

The second psychological argument is that loud sounds hide softer sounds. For years we have been told of the 'Cocktail-party effect', in which the ear picks up softer conversations over a loud babble, but this now seems to be discarded as far as music is concerned. It is argued that the overall loudness of a piece of music allows softer sounds to be neglected – in effect this means that the threshold of hearing can be adjusted according to the overall amplitude, dispensing with the need to cater for the lower amplitudes in such a mixture.

This latter argument seems to me to be on shaky ground. When a full orchestra is playing, can we ignore one oboe? If so, why not both? Are the bassoons required? Are the clarinets really heard? Conductors can certainly hear instruments whose contributions seem to be masked by the volume of sound, and other listeners might also find that the disappearance of 'insignificant' contributions made a noticeable change to sound quality. For many purposes, the saving in bandwidth might override the effects on the sound, but my feeling is that the full-bandwidth CD will still be the preferred medium for distribution of music for which high-quality recording is important. This point is particularly applicable to the Sony Mini Disc system (see page 126), which some fear might replace CD.

The third factor that allows reduction in bandwidth is the 'bit-bank' use. Each digital number used in the coding of an analogue signal can make use of a maximum number of bits (the bit-bank), but never does so because only a white-noise signal ever occupies all of the frequency spectrum and requires each bit of a digital signal to be used. Normal music only ever uses a fraction of the total bits in each sampled unit, and the spare bits (which will not necessarily be the same bits in each sample) can be used for other purposes.

By taking advantage of all three of these factors, it is possible to reduce the bit rate of a recording to less than one quarter of the rate used for a CD recording. Philips refer to this processing as PASC, Precision Adaptive Sub-band Coding, and has devised ICs to carry out the complex processing that is required, some aspects of which will sound familiar to anyone who has encountered the later versions of Dolby processing.

The audio frequency signal, following sampling into digital form, is split into 32 sub-bands, each with the same bandwidth. The important point here is that digital filters are being used. Such filtering would be impossibly elaborate with the analogue signals, and could not achieve the separation between sub-bands that is possible with digital filters. For each sub-band, the signals are passed only if they are above threshold level, and the threshold level is varied according to the overall amplitude (a dynamic threshold level). This is done by applying a set of numbers obtained from studies of the human ear and held in the memory of the processor. Since these numbers are stored in the processing chip, they can be altered by replacing the chip in order to modify the characteristics of the system if this should be required at any stage.

The use of the dynamic threshold filtering will alter the number of bits that each sub-band needs. A sub-band that makes use of only a few bits will be allocated what it requires, allowing other sub-bands to use more. In this way, a comparatively small number of bits can be more efficiently used, presenting the same overall effect as a system using a much larger number of bits. It is essential, of course, for the system to keep track of which bits are allocated for which band.

Once this coding has been completed, the bits are gathered, along with the usual Reed-Solomon error detecting and correcting codes, into eight channels of bits. The way in which bits in each channel have been allocated performs an action similar to that of interleaving, so that further interleaving is not needed. A ninth channel is devoted to control and display signals (and probably will have spare capacity). For each channel, eight-to-ten modulation is used to avoid long runs of 0s or 1s. These nine channels are then recorded on to the tape, using a compact cassette format with chrome tape of videotape standard.

Mechanical features

The form of the cassette (Figure 7.3) is certainly not identical to the older type, though of the same dimensions. In particular, the DCC has holes for tape drive on one side only, and is intended for use only in systems using auto-reverse. This automatically excludes it from a large number of existing players (particularly the high-quality ones), so that the market for the older type of cassette is not likely to suffer too soon. Philips believe that all cassette players will eventually use auto-reverse, eliminating the need to turn over a cassette.

The other striking difference is that the cassette has sliding covers for both tape and drive holes, using the methods pioneered

Figure 7.3 Digital compact cassette.

by computer mini-discs and also by videotapes. This make a holder unnecessary, and with one side of the cassette flat and free of holes, the label can contain more information than was possible on the older analogue cassette. Blank cassettes will have a tape length indicator hole and a record-protect arrangement.

The head design allows both digital and analogue tapes to be played (though not necessarily permitting analogue recording). The head can be rotated as part of the autoreverse action, and in each position, half of the head will be used for nine digital channels, the other half for two analogue channels. The analogue channels conform exactly to existing cassette standards, and when the head is turned, the positions of digital and analogue sections are reversed. The head is constructed using thin-film wafers, a technique developed for S-DAT.

Of the digital tracks, one is devoted to control and display information, and not all players will necessarily make use of this fully. Each digital track is 185 μm wide, but of this only 70 μm is needed for playback so that the tolerance of track width and for alignment of the tape is comparable with that of a normal cassette. The use of a reduced set of bits and eight channels of signal allows the shortest recorded wavelength to be 0.99 μm, easily accommodated on chrome tape, and avoiding the need for pure iron tapes as are needed for 8mm videotapes and DAT tapes.

Performance

The performance that can be achieved is impressive on paper, allowing sampling frequencies of 32kHz, 44.1kHz or 48kHz with corresponding frequency upper limits of 14.5kHz, 20kHz and 22kHz respectively. The lower limit of frequency is quoted as 5Hz. The dynamic range, as would be expected of a digital system is high, 105dB. The audio bit rate is 384kbits per second, and the recording time is the usual C90 45 minutes per side — there is provision for the use of 2 x 60 minute cassettes. These cassettes will be designated as D45 and D60 respectively.

The Digital Mini Disc

While Philips have concentrated on DCC (their agreement with Sony allowed access to Mini Disc technology, but they decided against it), Sony have been working on a radically different approach. The interaction between light and magnetism has been known since the 1820s (the Faraday effect), but only the development of small lasers and new magnetic material has made the effect one that could be used for low-cost read-write devices. A magnetic field will alter the plane of polarization of light, quite a different effect from the scattering that takes place at each pit of a CD. Since this effect is probably less-well known, a description appears in Appendix 3.

Mechanics

The basis of the Mini Disc (MD) technology is the reduction of the number of bits needed for coding, as described above for DCC and using the same basic ideas. The system used by Sony is called ATRAC (Adaptive Transform Acoustic Coding), and the reduction used for MD is more severe, resulting in only one fifth of the current CD rate being required. This allows the use of a 2.5″ diameter disc to store up to 74 minutes of music. The disc is in a protective cover with sliding shutters and the whole assembly is half the weight of an analogue cassette (Figure 7.4). The standard Serial Copy Management System (SCMS) is used, as for DAT, to prevent a copy being made from a copy to satisfy the requirements of copyright, allowing the user to make a copy but not to use such a copy to make others.

Unlike any other disc medium, however, MD offers both read and write actions, and MD players are geared to cope with both. The methods that have been proposed are curiously mixed. The magneto-optical methods are ideally suited to recording and

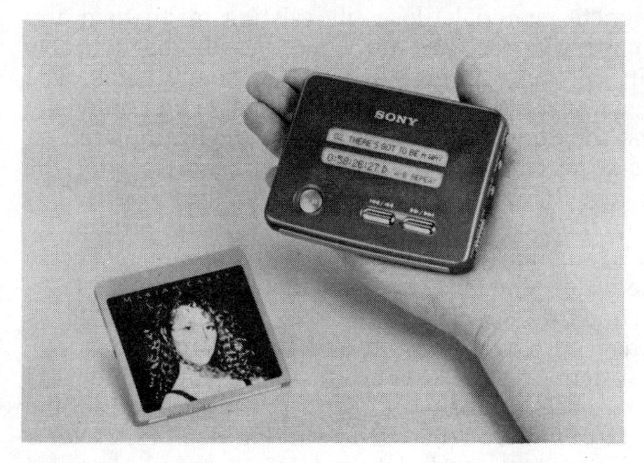

Figure 7.4 Sony's Mini Disc.

playback by the user, allowing the equivalent of cassette use for recording. For pre-recorded discs, however, this method is not well suited to mass production, so it is assumed that pre-recorded discs would be manufactured by the well-established methods used for CDs, and making use of the same equipment. This cleverly avoids the need to set up a whole new production line for these discs.

The effect on a laser-beam is not the same however, and MD players will incorporate two sets of detectors, one for the scattering effect of the conventional type of pits on the beam, the other for the polarized light signals from the surface of the magnetized discs. The correct detector can be switched into use automatically by detecting the magnetization of the disc surface.

Magneto-optical system

Magneto-optical recording and replay are not new, but in the past it was impossible to devise systems that could be manufactured at low cost, and the system has been used mainly for computer storage systems. In particular, it has always been difficult to erase old information in order to record new information, and Sony has solved this in order to make the whole process much simpler.

The basis of magneto-optical recording is the use of a magnetic coating on the disc which uses unusual materials, chemically classified as 'rare-earth elements'. The magnetic characteristics of these materials are unique — at normal temperatures they have very high coercivity so that their magnetism can be changed only by very strong fields. When the material is heated, however, it (like other magnetic materials) loses its existing magnetism and (unlike other materials) becomes much more easily magnetizable and will retain an imposed magnetic field when it cools. The heating of the material can be carried out by using a high-intensity laser beam, and the Sony system uses a magnetic head placed on the opposite side of the disc from the laser beam to alter the magnetization just following the instant when the material has been heated by the beam. This allows discs to be written by magnetic signals, and it make it unnecessary for any separate erase signal to be used, since the heat generated by the laser beam erases any old signals and the material has cooled to the extent that allows it to be remagnetized when it passed over the recording head.

The pickup system uses the laser beam at much lower intensity, together with two sets of analysers. For a pre-recorded disc, the light from the laser will be reflected to varying degrees, as for a conventional CD system, and detected by the same form of photodiode array. For a magnetically recorded disc, the light will be reflected with an angle of polarization that will differ according to whether a 0 or a 1 is recorded, and the detector uses polarising filters and photodiodes. The direction of polarization of the reflected light will determine which of the photodiodes generates more current, and by subtracting the signals, this will yield a 0 signal for one direction of polarization and a 1 signal for the other direction. In either case, the signals will then be processed in the usual way, but with the adaptations required for a reduced bit-number form of signal, as outlined for CCD.

Use on move

Another novel feature of the Sony system allows the MD to be used in 'Walkman' applications. The pickup can read data from the disk (either coding system) at 1.4Mbit per second, but the decoder can operate at only 0.3Mbit per second. This allows the use of a memory store which will provide bits if jolting the player has moved the pickup position. When the player starts, bits are placed at full speed into the memory and are read out at the lower rate. The use of the pickup is suspended when the memory fills, so that the pickup is being used intermittently. If a severe jolt disturbs the pickup position, the amount of data in the memory allows for three seconds of playing, more than enough for the pickup to be returned to its correct position. The repositioning is done by temporarily recording pickup position every 13 milliseconds and using this to return the pickup to its position if it is moved.

Specifications

The specifications for the MD system are similar to those for DCC. The sampling rate, however, is fixed at 44.1kHz, and the bandwidth at 5Hz to 20kHz. The dynamic range is 105dB and wow and flutter are unmeasurable. The disc speed (linear) is quoted as 1.2 to 1.4 metres per second, allowing a playing time of up to 74 minutes.

The future of DCC

Sony envisage the MD format as a replacement for the cassette, placing it into head-on competition with the DCC for this market. In this respect, the compatibility feature of DCC might prove to be irresistible for many audio enthusiasts who have already suffered the trauma of re-building a collection of vinyl discs. One fear that has been expressed is that record companies might find that they could produced pre-recorded MD (which are conventional plastic with no magnetic coating) cheaply enough to make them a rival to CD, eventually driving out CD and leaving no medium for distributing music of full range, free of bit-reducing technology. It appears the Philips and Sony will allow the consumers to decide, so that we might see a rerun of the battle of standards that bedevilled videotape recording for so long. Where this would leave DAT and other rotating head technologies is much easier to decide – there appears to be virtually no place other than as a high-cost system for a few users.

The Denon DTR-2000

The Denon DAT recorder-player (Figure 7.5) was the first domestic machine to be extensively marketed in the UK, though it was followed closely by a machine from Aiwa. The machine permits recording from Optical (CD), coaxial (other digital) or analogue signal sources, and includes a digital fade-in and fade-out action for all inputs. The use of the sub-codes allows a large number of control features to be offered, such as Autostart ID to identify the start of each item, Skip ID to omit parts, Manual ID to insert a code, End ID to mark the end of a recording and Auto-renumber which will renumber all of the programs in order.

One point about a DAT system is that it must include both A-D as well as D-A stages, because it will be necessary to provide for analogue inputs that will be converted to digital for recording on the tape. The Denon uses sigma-delta modulation (see Chapter 4) for its A-D stage, allowing a high sampling rate to be used. The D-A stage is Denon's Lambda SLC type (see Chapter 4), rather than any form of bitstream.

Figure 7.5 The Denon DTR–2000.

The high-speed wind is at 400 times the normal tape speed, making the location of ID marks very rapid. There is also a Fine-Cue function which slows the tape to half-speed for precisely placing an ID marker. The machine offers three sampling frequencies of 32kHz, 44.1kHz and 48kHz. An amorphous head construction is used, and the claimed response is 2Hz to 22kHz ±0.5 dB, with a S/N of 90 dB. Total harmonic distortion is 0.008% and wow and flutter is unmeasurably low.

8 Sound synthesis

New sounds

Quite early in the history of electronics, it was realized that audio systems could become the basis for the generation of music rather than simply its reproduction. Almost as early as triode valves became available, the Theremin was demonstrated. This consisted of a valve beat-frequency oscillator system coupled to a loudspeaker and amplifier. By waving his hands near a metal bar, the player could alter the frequency of one oscillator, so that the beat note, which was normally almost zero, changed so as to give a 'musical' note. Early devices like the Theremin were exhibited widely, and the principles of synthesized music were established long before there was any significant demand for its use. The real start of synthesized music, however, begins with the second series of Hammond organs, introduced in 1939, which used valve oscillators as the source of audio tones which could be switched by a keyboard to an output amplifier and loudspeakers. The synthesizer, so long regarded as the province of the pop-musician, thus had its roots in an instrument that was very widely used in church halls and also in churches.

Synthesizers in general use two routes to the final sound. One route is the voltage controlled oscillator, as pioneered by Moog, the other is the noise source with controllable filters. The voltage-controlled oscillator is the logical development of the Theremin principle, and uses an oscillator or set of oscillators which can have the frequency controlled by a DC voltage applied from voltage dividers. Each key of a keyboard instrument operating on this principle therefore need only operate a simple DC voltage divider rather than alter a frequency-determining component such as a

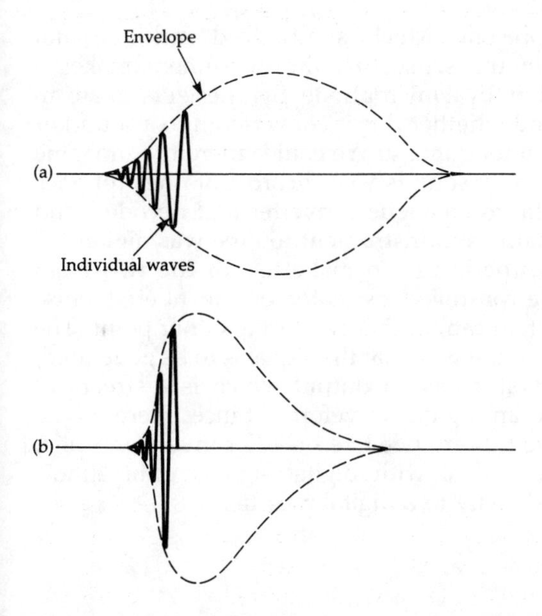

Figure 8.1 Wave envelopes, which are the shape formed by the peak amplitudes of the waves in a note. The sound of a note is affected by both the shape of the waves and also by the shape of the envelope.

capacitor or inductor. The noise-source method starts with a generator of wide-band noise, and uses the keys to control filters which pass a narrow bandwidth. In this case, the action of the key is more complex, and often consists of gating the noise waveform to an active filter.

The two methods produce very different sounds. The voltage controlled oscillator can produce a waveshape that depends on how the oscillator is biased, so that shapes ranging from sine to square can be produced. The noise filter method inherently produces a more complex waveform, but in some cases not a particularly musical note. Both systems are capable of an almost infinite variation of effect, and can have envelope shapes (Figure 8.1) imposed which will determine whether the note sounds like a specific musical instrument or something quite different.

Both methods are analogue, and the introduction of digital methods into synthesis has taken place step by step, and is not yet complete. This, however, is one field in music where digital methods could eventually be totally dominant, with no analogue steps of any kind present. Much synthesized music is not 'performed' as such in the sense that the instrument makes a sound which is picked up by a microphone. Synthesizers create an electrical waveform, and whether this is converted into a sound or not is just one option among many. We could, therefore, imagine a system in which the only sounds were heard from headphones run from a small digital to analogue converter and decoder, and the whole recording path from instrument to tape was digital.

Just as analogue synthesizers can make use of the two main sources of the voltage-controlled oscillator or the filtered noise source, so a digital system can, in theory, start at either point. The obvious difference lies in the way that the signal is to be generated, because a digital generator has an output which is a stream of numbers rather than an audio waveform. Since there is no 'original' signal involved, there need be no AD conversion stage, and the equipment can deal with digital signals until almost loudspeaker level, or directly to a digital recorder.

Synthesis methods

A digital audio generator consists of a unit in which a series of numbers can be generated whose value corresponds to the amplitude samples of an audio wave. The simple and obvious way to do this, of course, is to use a voltage-controlled oscillator along with an AD converter, but this is an analogue conversion rather than a true digital generator. The alternatives are the use of a microprocessor system to generate numbers that follow a mathematical pattern (simple like that of a sinewave or complex like a sine/cosine series), or the use of ROMs that contain sequences of numbers that can be used to build up waves of specified shapes. The use of ROMs can be very similar to the way that tape has been used in synthesizers to hold a basic sound pattern which can then be altered by the action of the synthesizer.

The other starting point, of using a noise signal, is less promising. Though a random (or pseudo-random) number generator can be used to provide a digital 'noise' signal, digital methods that provide the equivalent of filtering are not quite so simple. A low-pass filter, for example, implies that numbers are reduced on the basis of the difference between numbers, the equivalent of smoothing steeply rising or falling waveforms. Digital noise generators are already being used to some extent, but coupled to DA converters and analogue filters to achieve synthesis by this type of action. Digital filters are, however, available and their use is increasing.

The main extent of the penetration of digital techniques into synthesis has been concerned with time delays and control techniques. The introduction of controlled reverberation into synthesized signals was at one time carried out only the tape loops and acoustic delays, but units based on charge-coupled devices (CCD's) have become more common as the price of IC's has fallen. The next logical step to fully digital delay systems has not been long delayed, and the rapid fall in prices that has affected all digital devices from watches to computers has made its mark here too.

The action of a digital delay circuit for existing analogue signals makes use of sampling and an AD converter in order to transform the analogue signal into digital form. The digital samples are then stored in memory until the time comes to read them. At one time, serial memory (a shift register) would have been used, but modern techniques demand such luxuries as variable delay time and programmable delay time, so that random access is used greater extent now. The sampling rate is comparatively low on the more modestly priced delay units, and only the top-price units use sampling rates that approach the 44.1 kHz of the CD system. Such units can offer delays of up to a minute that require huge memory capacity – the use of high-speed magnetic discs like the computer hard-disc is often a more useful technique than solid-state memory for such long term delays.

MIDI action

The most noticeable impact of digital methods upon sound synthesis has been the adoption of the MIDI (musical instrument digital interface) standard both by the manufacturers of musical instruments and the manufacturers of some home computers. The requirement for some standard arose from the increasingly complicated setting procedures that synthesizers required. Since it would be unacceptable to take an interval of a minute or so between notes in order to reset a machine, the principle of using a microprocessor to control the settings has been very attractive for the users of synthesizers. This was done, notably on the Roland Jupiter 8 and the Sequential Circuits Prophet 5, but as other manufacturers rushed to devise sequential microprocessor controls for their own equipment, the lack of standardization became a handicap. It was soon obvious that without standardization, users of microprocessor-based equipment would be confined to the products of manufacturers who made a full range of equipment that used the same methods.

The system, originally proposed by Dave Smith of Sequential Circuits, was modified in the light of suggestions from Roland, Yamaha and others to emerge as the MIDI standard. Though this uses a serial interface with the inevitable limitations of speed, the interface is very cheap to implement and cable length is less of a problem than is the case when parallel-signal systems are used. The standard speed of 31.25 kbaud is well above the speeds used by most computer serial links, and is adequate for most purposes. In this respect, one baud means one signal unit per second and though baud rate does not necessarily mean bits per second, it is usually regarded as having this meaning.

The MIDI system allows codes for notes that span 10 1/2 octaves, and also permits the signalling of changes of memory or for information that is specific to an instrument. The system, though originally devised for controlling keyboard synthesizers can also be used for timing drum and percussion synthesizers, and as a result, MIDI interfaces can be found fitted to a considerable variety of devices – including several models of home computers, notably the Atari ST models.

The MIDI system allows up to 16 separate channels to be used, so that the normal fitting of a MIDI interface allows for signals both in and out, except for a device that originates or receives signals only. The system uses 8-bit groups (bytes) of data, and in each byte the individual bits will carry their own information. For example, the channel number can be carried by four bits of one byte, and in this same byte another bit can indicate the type of byte, which would be a status byte if it carried channel number codes and whether or not the following bytes refer to note information. The decoder thus can separate a byte that is used for channel selection from one which is used to carry tone information.

These MIDI codes allow information on note (pitch), volume level on each note, and the type of pitch and amplitude envelopes of each note to be sent both between instruments and from a computer to an instrument. The instruments that are used in a modern MIDI link can operate in any one of four different modes, replacing the older three-mode system (termed omni, poly and mono), and the mode selection governs the way that the transmitted signals affect the individual instruments.

Mode 1 is the most basic, and MIDI systems in general will run in this mode at switch-on. This mode ignores MIDI channel information so that the equipment will respond to signals on any channel – the instrument will try to play each note that is received in turn, no matter on what channel. The receiving instrument is assumed to be polyphonic – capable of playing more than one note at a time. This mode was incorporated with the needs of beginners in mind and is intended for a very simple system in which a master instrument is connected to a single slave instrument.

Mode 2 is one that is not used to any great extent, because it assumes that the receiving instrument is monophonic, one note at a time, an obsolescent breed. Using Mode 2 with a polyphonic master instrument risks causing confusion because the receiving instrument will not be able to cope with the number of note codes; the usual solution is for the receiving machine to select the last-received note, or the lowest or highest pitch, depending on settings.

In Mode 3, each instrument is allocated its own channel out of the sixteen that are available, and note information is ignored unless it is on a specified channel. This allows the transmitter to take individual control over each instrument so that each can play an individual part. The notes are, of course, actually sounded in sequence, but because of the comparatively high rate of the MIDI transmission the delay is not audible, and even with sixteen different parts being played, the sounds appear to be simultaneous. The MIDI system needs to be set up so as to ensure that the same channels are being used by both transmitter and receivers.

Mode 4 allows for the use of modern synthesizers which can play chords. The older type of synthesizer had one 'voice' meaning that only one note could be played at a time, though special chord keys could be used to provide harmony like the chord buttons on a piano-accordion. More modern units allow multi-voice operation, so that chords can be played in the same way as they are on a piano. This allows much more control over the type of harmonies, since the fixed chord scheme allows only the main major and minor chords to be used.

Mode 4 uses channel 1 of the MIDI transmission to determine the note and other features of voice 1 of the synthesizer, and the other voices are controlled by the other channels. In addition, this mode allows the control of features which may be unique to an instrument, like variation of key pressure affecting loudness, or in the reproduction of glissando. The snag is the many expensive instruments which could exploit mode 4 very effectively are not equipped to do so.

Appendix 1
Further reading

The following is a selection of useful books and magazines which can be used in conjunction with this book and for further reading.

Books

Digital theory and methods

Digital Logic Gates and Flip-flops by I.R. Sinclair, published by PC Publishing 1989, reprinted 1990. A guide to the use and design of digital gate circuits for the electronics enthusiast.
Digital Interfacing with an Analogue World by J. Carr, published by TAB 1987. Deals with analogue to digital conversion and interfacing generally.
Digital System Design by Barry Wilkinson, published by Prentice-Hall 1987. An introduction to the design of digital circuits with emphasis on the use of microprocessors.
Electronic Filter Design Handbook by Williams & Taylor, published by McGraw-Hill 1988. A specialized book which is one of the few that deals with digital filter theory and design.

Audio, general

Audio Electronics Reference Book edited by Ian Sinclair, published by Blackwell Scientific 1989. A large reference book on all aspects of audio technology with contributions from well-known specialists.
Magnetic Recording Handbook by Mee & Daniel, 1990, published by McGraw-Hill. A very large manual on all aspects of magnetic recording, both analogue and digital.

How to Set Up a Home Recording Studio by David Mellor, PC Publishing 1990. 124 pages, 40 line drawings, 35 photos. A complete guide to building an 8 to 16 track recording studio.

Music and MIDI

Practical MIDI Handbook 2nd edn, by the well-known R. A. Penfold, published by PC Publishing, 1990. 176 pages, 37 line drawings, 7 photos. A full introduction to and explanation of MIDI modes and channels.
Advanced MIDI User's Guide also by R. A. Penfold and published by PC Publishing, 1991. 160 pages, illustrated. A guide for the experienced MIDI user.
Computers and Music, by R. A. Penfold, published by PC Publishing 1989. 176 pages, 50 line drawings 15 photos. Covers the basics of using a computer, via the MIDI system, to control electronic musical instruments.

Digital audio

Digital Audio and Compact Disc Technology is by the staff of the Sony Service Centre (Europe), notably Luc Baert, Luc Theunissen and Guido Vergult, published by Heinemann Newnes. It contains a very full and well-illustrated account of digital audio systems (though not to the point of mentioning bitstream) and is particularly useful to the reader who already has a good grasp of digital methods.
The Art of Digital Video by J. Watkinson, published by Focal Press 1990. Deals with the techniques of digital audio and video recording by both magnetic and optical media. Treatment is non-mathematical.
Principles of Digital Audio by K. C. Pohlmann, published by Howard Sams, 1989. Deals with all aspects of digital recording and replay.
Advanced Digital Communications edited by Dr. K. Fehrer, published by Prentice-Hall 1987. A massive reference book dealing with all aspects of digital conversion, communication and recording.

Servicing and fault diagnosis, general

Understanding Digital Troubleshooting by the staff of Texas Instruments, published by Howard Sams 1984. A guide to methods of locating and curing faults in digital circuits.

Advanced Digital Troubleshooting by A.J. Evans, published by Howard Sams, 1988. A guide to modern methods of diagnosis of faults in digital equipment.
Television and Video Engineer's Pocketbook by Eugene Trundle, published by Heinemann Newnes, 1987. An excellent introductory text for anyone who has lost touch with the developments in video over the last few years.

Magazines

The audio magazines are by far the best way of keeping abreast of the rapid developments in audio technology. The following list is far from comprehensive, but it includes magazines which have been particularly active in presenting information on developments in digital audio techniques.

Hi-Fi News and Record Review
What Hi-Fi?
Hi-Fi Choice
Digital News (Philips)

Appendix 2
Glossary of terms

A-D
Analogue to digital conversion

Address
A location on a magnetic or optical disc or in a memory. Each address location is identified by its address number, usually referred to as the **address**.

Aliasing
The generation of spurious frequencies caused by using a sampling rate that is less than twice the highest frequency of signal present. For example, if a 15 kHz signal is sampled at 25 kHz, there will be a strong 5 kHz component, equal to the difference between 25 kHz and 2 × 15 kHz.

Analogue
Using an infinite number of signal levels between maximum and minimum (usually zero).

Asynchronous
Not tied to a fixed rate of repetition. An asynchronous signal can occur at intervals which do not coincide with a fixed-rate clock pulse.

ATF
Automatic track following, the system used in the R-DAT player to ensure that the rotary heads follow the recorded track. This uses a set of signals that is recorded along with the digital data and which are passed to the servo controls to ensure that the tape is correctly positioned with respect to the heads.

Audio
Referring to the frequency range of human hearing, in the range of 30 Hz to 20 kHz.

Azimuth
The angle of the slit in a tape-head to the line of a perpendicular drawn to the tape edge, normally zero. A difference in azimuth between recording and replaying heads will cause a large loss of signal amplitude. This is used deliberately in rotating-head machines to ensure that interference between adjacent tracks is minimised.

Bandpass
A filter circuit which passes a range of frequencies, rejecting frequencies below the lower limit of the band and above the upper limit. A 25Hz to 15kHz bandpass is often applied to audio signals.

Bias
An additional voltage or signal applied so as to use the linear part of a device or recording system. Voltage or current bias is required for semiconductors, AC signal bias for analogue magnetic recording systems. Digital systems require no biasing.

Binary
Using only two digits, levels or codes, as opposed to the scale of ten used in normal counting.

Bit
A binary digit, 0 or 1. The position of a bit in a binary number determines its significance (its value as a power of 2).

Bitstream
A technique pioneered by Philips in which a binary number is represented by a set of signals of two possible levels. The average value of these signals corresponds to the analogue voltage which was encoded by the binary number.

Byte
A set of 8 digital bits, the normal unit of memory used in small computers.

CD
Compact disc, a recording made using optical markings on a plastic

disc, formed and also read by a laser beam. The markings are digitally coded to represent samples of the amplitude of sound waves.

Checksum
A number derived from arithmetical actions on data and used to check that the data has not been corrupted after transmission or recording and replay.

Clock
A pulse which is repeated at a steady rate in order to control the timing of actions. The clock pulses are usually derived from crystal-controlled oscillators, and rates of up to 33 MHz are used in computing circuits.

CMOS
Acronym of complementary metal oxide semiconductor, applied to ICs which use both N-channel and P-channel MOSFET technology in device pairs

Combinational circuit
A circuit whose output depends on the combination of inputs that is present at any instant. Also known as a logic or gating circuit.

Counter
A circuit whose output depends on the number of pulses at a input, so that the state of the output can be used as a number count.

CRC
Cyclic redundancy check, a system of recording a checksum number along with data in order to detect and in some cases correct any corruption of the data.

D-A
Digital to analogue conversion, the essential transformation at the player which permits the sound wave to be reconstructed from the digital number data.

DASH
Digital audio stationary head, a system of recording digitally-encoded sound using a large number of parallel tracks on tape.

DAT
Digital audio tape, any system of recording digitally coded sound onto tape, using either multiple heads and fast tape speeds (S-DAT) or revolving heads and slow tape speeds (R-DAT).

Demultiplexer
A circuit which separates combined streams of data into separate streams.

Differentiator
A circuit whose output represents changes in signal level only, with no response to steady signal levels.

Digital
Using only two digital levels, labelled as 0 and 1.

Dither
Addition of random signal (noise) in order to avoid perfectly steady signal conditions.

Dropout
A tape fault in which a small piece of magnetic materials is missing or faulty, causing a loss of signal at that point.

Dynamic memory
A form of memory which depends on storing charge in small IC capacitors. Because of leakage, the stored charge needs to be refreshed at intervals of a few milliseconds. Dynamic memories can be built in very large sizes, of the order of 1M – 4M bits per chip.

EFM
Eight to fourteen modulation, a method of recoding a byte of data so that the 0s and 1s fall into a pattern that is more suitable for recording.

Electrostatic
Depending on electric charge. Any objects rubbing against each other will be electrostatically charged, and if both are insulators, the voltage levels obtained in this way can be very large, of the order of several kilovolts.

EPROM
Electrically programmable read-only memory, a memory chip which can be programmed by electrical signals, see PROM.

Excess-3 Code
A form of binary code in which 3 is added to each number before binary coding. In this code, there is no zero number (0 codes as 0011), some arithmetic actions are simpler, and the sum of a single-digit number and its complement is always 9.

Fanout
The ability of a digital IC to drive others, expressed as the number of inputs that can be driven from a single output.

FET
Field-effect transistor, a switching and amplifying device that controls current flow in a ribbon of semiconductor by means of an electric field. See also MOSFET.

Field
In TV, one half-set of picture scan lines, either the odd-numbered lines or the even-numbered lines.

Filter
A circuit which will pass one range of frequencies, the passband, and reject all others (the stopbands). See also lowpass, highpass, bandpass.

Flip-flop
A circuit whose output can be switched to either state (0 or 1) by an input.

Frame
In TV, one complete set of picture scan lines, composed of two fields.

Frequency
Number of pulses or repetitions of a waveform per second.

Gate
A digital circuit whose output depends on the combination of inputs that are present.

Glitch
In general, any false digital signal. In digital audio, a transient that occurs mainly in the D-A conversion which can be removed in a low-pass filter.

Gray code
A system of coding numbers using 0 and 1 digits, but in which the change from one number to another is always small, one digit only, as a count is made.

High pass
A filter circuit which passes only frequencies above a limit, rejecting lower frequencies.

Integrator
A circuit which has little or no response to sudden changes in input, but smooths the input into a steady or slowly-changing output.

Interpolation
Addition of information by taking an average value between two adjacent items.

Lambda SLC
A technique for digital to analogue conversion, due to Denon, and meaning 'ladder-form multiple bias D/A super linear converter'. This uses 8 times oversampling to produce a 20-bit signal which is split into two streams to reduce the effect of cross-over biasing.

Laser
Acronym of 'light amplification by stimulated emission of radiation'. A device which for an electrical input generates light of a single wavelength and with no sudden changes of phase.

Linear
Any system in which a graph of output plotted against input produces a straight line.

Logic
A system for producing answers from questions. In digital circuits, a logic circuit will produce an output when the inputs are in a predetermined pattern.

Lowpass
A filter circuit which passes only lower frequencies, rejecting frequencies above the limit. A 15 kHz lowpass filter is often used in audio work to attenuate or reject higher frequencies which could cause trouble in circuits.

MASH
The name used for the Technics multi-stage noise shaping circuits, a form of bitstream D-A conversion.

Memory
A circuit which retains one of two possible states once set. Memory can be volatile or non-volatile.

MFM
Modified frequency modulation, a method of coding binary data, particularly for computer magnetic discs, in which a 1 is coded by a transition (change of direction of magnetization), isolated zeros are ignored and a transition is placed between pairs of zeros. This reduces the DC content of the signal as compared to NRZ coding. MFM has been superseded by EFM for digital audio work.

MOSFET
Metal-oxide-semiconductor field effect transistor, a form of FET (q.v.) in which the control is applied to a gate electrode which is isolated from the semiconductor by a thin film of silicon oxide. No current is taken by the gate, so that MOSFETs can operate at very low power levels. In addition, the construction is particularly easy to use in integrated circuits.

Microprocessor
An IC which carries out digital actions such as logic gating, counting and shifting in response to inputs of instructions (the program) and data.

MIDI
Acronym of 'musical instrument digital interface', an agreed standard for linking electronic musical instruments to each other and to computers in order to control several instruments at once.

Modulation
The alteration of signals into a form that can be transmitted or recorded.

Monochrome
Of one single frequency, like the light from a laser. Also applied to black/white TV.

Multiplex
Any system of encoding more than one signal on a line. This includes frequency-multiplex, using several carriers at one, combined phase and amplitude modulation as used for colour TV, and time-multiplex, in which the signals for different channels are sent at different parts of a timing cycle.

NMOS
Acronym of 'negative-channel metal-oxide-semiconductor'. One of the three possible forms of MOSFET ICs, the others being PMOS and CMOS. The NMOS IC uses predominantly N-channel MOSFET technology.

Noise
Unwanted signal of any kind, usually of random frequency and amplitude.

Noise immunity
A measure, usually in terms of volts, of the ability of a digital circuit to ignore noise signals.

Noise shaper
A circuit, used in bitstream D-A converters, whose action is to shift the frequency of noise in a digital signal so that on conversion the noise will be outside the audio range. This is done by reducing the number of parallel bits used to carry the data, increasing the number of serial bits and so the frequency of the digital signal.

Non-volatile
Applied to memory that retains information when its power supply is switched off. This is true of all magnetic memory, and also of ROM and some types of CMOS RAM.

NRZ
Non-return to zero, the conventional digital magnetic recording system in which magnetization in one direction represents logic 1 and magnetization in the opposite direction represents logic 0.

NTSC
National Television Standards Committee, the US body which agreed the original compatible colour TV standards in 1952.

Optical
Making use of light, including infra-red and ultra-violet frequencies which are invisible to the human eye.

Oversampling
A method of increasing the performance of a D-A converter by inserting additional pulses into the digital signal between the original pulses. The effect is similar to that which would be produced by using a faster sampling rate in the original recording.

Parity
A system of checking data by recording an extra bit in each byte. The extra bit is used to check the number of 1s in the byte by determining if this number is odd or even.

PCM
Pulse-code modulation, a system of converting analogue signals to digital in which the amplitude of the analogue signals is converted to a binary number, rather than to pulse amplitude, pulse frequency or pulse position. All digital audio systems use PCM.

Photoresist
A material which is used in the manufacture of printed circuit boards, ICs and compact discs. The photoresist is normally soluble in an alkaline solution, but it becomes insoluble when exposed to light, so that a surface which has been exposed through a photographic negative can be 'developed' in an alkaline solution to leave a pattern of hardened resist. The exposed surface that is left can then be chemically etched, leaving a pattern which will remain when the remaining photoresist is removed.

PROM
Programmable read-only memory, an IC which contains data that cannot be lost when the power-supply is turned off. In its most familiar form (EPROM), the PROM can be programmed by using signals of higher level than normal, and these signals will be retained until the PROM is 'washed' by exposure to ultra-violet light.

Pulse
A short-duration change of electrical voltage or current from one level to another and then back again.

Pulse code modulation
see PCM, above.

Quantization
Analysis of a waveform into a number of set levels, the essential preliminary to any analogue to digital conversion.

Quantum
A unit of anything. Originally applied to the quantity action (energy × time) in Planck's Quantum Theory, and used in digital work to describe a unit of signal level.

Quartz crystal
A crystal of quartz cut and with metal film deposited as as to be used in an oscillator whose frequency will be precisely set by the dimensions of the crystal.

Race hazard
A problem of gate circuits, in which spurious signals are formed because the inputs to a gate arrive at slightly differing times. Race hazards can be eliminated by ensuring that all signals are synchronized to a master clock signal.

RAM
Random access memory, used now to refer to read-write memory, since all modern memory systems use random access. RAM is often volatile so that its contents will disappear when power is switched off.

Random
Corresponding to no fixed pattern, such as numbers drawn from a hat.

R-DAT
Rotary-head digital audio tape, the system adopted for domestic DAT systems, using two rotating heads spinning in a plane that is slightly angled to the tape. This allows a low tape speed to be used along with a high head-to-tape speed; the same system as is used for video recording.

Red Book
A data book which contains the specifications for the CD and other digital audio systems, provided to manufacturers who take out a licence for using the system.

Redundancy
The addition of excess information so that the loss of some data can be made up by using the rest of the data.

Reed-Solomon
A coding system that uses redundancy to detect and correct errors in data transmission.

Refresh
The action of applying pulses to a dynamic memory chip to rewrite the 1 bits of data.

Register
A temporary store for digital data, in which data can also be manipulated using addition, subtraction and bit shifting.

ROM
Read only memory, a form of non-volatile memory which uses fixed connections in a chip to provide data that cannot be altered.

Rotary head
The video-recording system that uses a rotating drum carrying two or more heads which sweep across the tape at a small angle, typically 5° to 8°. This allows a high rate of scanning to be combined with a low tape speed.

Sampling
The system of measuring the amplitude of a waveform at fixed intervals so that the samples can be converted into digital signals.

Saturation
A state of complete magnetization of a magnetic material, in which no increase of magnetizing effort can produce more retained magnetism.

SCMS
Serial copy management system, a form of coding used on DAT

machines which prevents a digital tape from being used to create another tape and so mass-produce perfect copies.

S-DAT
Stationary head digital audio tape, another name for DASH (q.v.).

Sequential circuit
A circuit in which the output depends on the number of inputs rather than on their state at a given time. All counter circuits are sequential.

Shift register
An array of flip-flops connected so that the output bits are shifted from one flip-flop to the next when a clock pulse is received on all the units.

Sideband
A range of frequencies produced when a signal frequency is modulated or sampled.

Sidebeam
A part of the laser beam of a CD player, used to detect whether or not the main beam is aimed correctly. Two side-beams are used to ensure that the main beam is held in the centre of the track.

Significance
The number that the position of a digital represents – the '2' in 120, for example represents the number of tens, and the '1' represents the number of hundreds.

Smoothing
The conversion from a waveform with a sawtooth or spiky outline into a smooth waveform, a process of integration.

Splicing
Joining cut edges of tape so as to make an edited version in which the joins are undetectable.

Static memory
A form of memory in which data is held as long as power is supplied, but without any need to refresh the data.

Switching
The sudden change from one state to another, such as conducting to non-conducting or non-conducting to conducting.

Synchronous
Acting in time to a set of clock pulses which are delivered at regular intervals.

Truth table
A table of inputs and their corresponding outputs for a combinational circuit.

TTL
Transistor-transistor Logic, a form of construction used for early digital ICs, in which the inputs are to the multiple emitters of a transistor. The system is still used in modified form for logic circuits.

Ultrasonic
Using frequencies above the normal limit of audibility (about 20 kHz). Very high ultrasonic frequencies, extending to tens of megahertz, are used in some equipment.

Video
Relating to images, applied to image recording and transmission.

Volatile
Applied to memory systems, meaning that all data will be lost when the power supply is switched off.

Waveform
The graph of voltage or current plotted against time for an electrical signal.

White noise
A noise signal in which all frequencies are present to an equal extent, not weighted in any way (as pink noise would be).

Appendix 3 Polarization of light

Light is, like radio waves, an electromagnetic vibration which travels through space. The amplitude of the vibration is, like waves in water, at right angles to the motion of the waves, but unlike water, the direction of the amplitude can be any direction at right angles to the movement, allowing a full 360° of possible directions. We take the direction of vibration as being the direction of the electric field (the magnetic field is at 90° to this). Normal, unpolarized, light uses all of these directions, there is no preferred direction.

Some natural materials, some angles of reflection, and some man-made substances will filter out light so as to permit the vibration of the electrical signal to be in one preferred direction. This effect is called polarization, and these materials are polarizers. The most striking demonstration of polarization is that once light has been polarized, passing it through a second polarizer can reduce the light amplitude to zero if the polarizer is turned so that its angle of polarization is at 90° to the angle of polarization of the light. Polaroid sunglasses exhibit this effect very noticeably, and they are supplied so as to reduce the amount of light that is reflected from horizontal surfaces (which is polarized quite strongly) so reducing glare.

Polarization had already been discovered by 1678, but understanding of polarization as evidence of light waves was developed mainly in the period from 1800 onwards. Faraday's discovery was that a beam of polarized light could have its angle of polarization altered by passing through a magnetic field, and since then, materials have been discovered which allow the effect to be greatly intensified. The material used in the Sony MD recording discs is terbium ferrite cobalt, an exotic compound which will alter the polarizing angle of light which is reflected from it depending on the extent to which the material is magnetized. The reflected beam can be passed to photodiodes, each with a polarizing sheet in front of it, so that the amount of light that enters each photodiode will depend on the angle of polarization of the beam. By making magnetization in one direction represent 0 and magnetization in the other direction represent 1, the relative amount of beam at the two photodiodes will alter for each change in magnetization, allowing digital signals to be obtained.

Index

Introducing DIGITAL AUDIO

CD, DAT and Sampling

Updated second edition

Ian R Sinclair

★ For enthusiasts, technicians and students ★
★ Digital techniques explained non-mathematically ★
★ Covers CD and DAT ★
★ New sections on oversampling and bitstream methods ★
★ Glossary of terms ★

Digital audio involves methods and circuits that are totally alien to the technician or keen amateur who has previously worked with audio circuits. This book is intended to bridge the gap of understanding for the technician and enthusiast. The principles and methods are explained, but the mathematical background and theory are avoided other than to state the end product.

This second edition has been updated to include sections on oversampling methods and bitstream techniques. The opportunity has also been taken to add a glossary of technical terms.

Reviews of first edition

'Readable and informative' *Home & Studio Recording*

'Well worth a read . . . the writing is clear and unambiguous' *The Gramophone*

ISBN 1-870775-22-8

9 781870 775229

PC Publishing